SpringerBriefs in Cognitive Computation

About the Series

SpringerBriefs in Cognitive Computation are an exciting new series of slim high-quality publications of cutting-edge research and practical applications covering the whole spectrum of multi-disciplinary fields encompassed by the emerging discipline of Cognitive Computation. The Series aims to bridge the existing gap between life sciences, social sciences, engineering, physical and mathematical sciences, and humanities.

The broad scope of Cognitive Computation covers basic and applied work involving bio-inspired computational, theoretical, experimental and integrative accounts of all aspects of natural and artificial cognitive systems, including: perception, action, attention, learning and memory, decision making, language processing, communication, reasoning, problem solving, and consciousness.

More information about this series at http://www.springer.com/series/10374

Azlan Iqbal · Matej Guid · Simon Colton
Jana Krivec · Shazril Azman
Boshra Haghighi

The Digital Synaptic Neural Substrate

A New Approach to Computational Creativity

Springer

Azlan Iqbal
College of Information Technology
Universiti Tenaga Nasional
Kajang, Selangor
Malaysia

Matej Guid
Faculty of Computer and Information Science
University of Ljubljana
Ljubljana
Slovenia

Simon Colton
Department of Computing of Goldsmiths College
University of London
London
UK

Jana Krivec
Department of Intelligent Systems
Jozef Stefan Institute
Ljubljana
Slovenia

Shazril Azman
College of Graduate Studies
Universiti Tenaga Nasional
Kajang, Selangor
Malaysia

Boshra Haghighi
College of Graduate Studies
Universiti Tenaga Nasional
Kajang, Selangor
Malaysia

ISSN 2212-6023 ISSN 2212-6031 (electronic)
SpringerBriefs in Cognitive Computation
ISBN 978-3-319-28078-3 ISBN 978-3-319-28079-0 (eBook)
DOI 10.1007/978-3-319-28079-0

Library of Congress Control Number: 2015960221

Printed on acid-free paper

This Springer imprint is published by SpringerNature
The registered company is Springer International Publishing AG Switzerland

For Chesthetica
mundi maxime proficiebat automatic
latrunculorum problem compositor

Preface

This book was originally intended as a journal article. However, after a year of trying to get it published with a reputable journal, the editors and associate editors agreed on one thing. It was simply too long and needed to be shortened. This meant removing critical or even important aspects of the research work that might have affected its ability of being properly reproduced or adapted to other areas of research by others. Therefore, we decided against it and found our way here instead. We are grateful to Springer for providing us this alternative avenue. The content has therefore been expanded somewhat and modified to suit the style of a book. We hope this work will prove to be a seminal and valuable reference to researchers not only in artificial intelligence (AI) but also other areas in the decades to come.

Acknowledgments

This research was supported by the Ministry of Science, Technology and Innovation (MOSTI) in Malaysia under their *eScienceFund* research grant (01-02-03-SF0240). We would also like to sincerely thank Vlaicu Crisan who provided detailed feedback pertaining to the DSNS-generated compositions, and Cameron Browne for his various contributions to the project.

Contents

About the Authors

Azlan Iqbal received the B.Sc. and M.Sc. degrees in computer science from Universiti Putra Malaysia (2000 and 2001, respectively) and the Ph.D. degree in computer science (artificial intelligence) from the University of Malaya in 2009. He has been with the College of Information Technology, Universiti Tenaga Nasional since 2002, where he is senior lecturer. He is a member of the IEEE and AAAI and chief editor of the electronic Journal of Computer Science and Information Technology (eJCSIT). His research interests include computational aesthetics and computational creativity in games.

Matej Guid received his B.Sc. (2005) and Ph.D. (2010) degrees in computer science from the Faculty of Computer and Information Science at the University of Ljubljana, Slovenia. He is an assistant professor at the Artificial Intelligence Laboratory, University of Ljubljana. His research interests include heuristic search, computer game playing, automated explanation and tutoring systems, and argument-based machine learning. Chess has been one of his favorite hobbies since childhood. He was also a junior champion of Slovenia a couple of times and holds the title of FIDE master.

Simon Colton is Professor of computational creativity and EPSRC Leadership Fellow in the Department of Computing of Goldsmiths College, University of London, and previously reader in computational creativity at Imperial College London. He heads the Computational Creativity Group, which studies notions related to creativity in software. He has published over 120 papers on AI topics such as machine learning, constraint solving, computational creativity, evolutionary techniques, the philosophy of science, mathematical discovery, visual arts, and game design. He is the author of the programs HR and The Painting Fool.

Jana Krivec received her B.Sc. in psychology at the University of Ljubljana in 2004 and her Ph.D. in psychology (cognitive science) in 2013 at the same university. She works as a researcher at Jožef Stefan Institute since 2004. Her research interests include cognitive science combined with artificial intelligence, in particular, the game of chess. Notably, she is also a chess woman grand master and several times Slovenian champion.

Shazril Azman earned diploma in information technology (IT) in 2009 and bachelor's degree in graphic and multimedia in 2012. He has been involved in projects and workshops related to Web development, graphic design, computer animation, and augmented reality throughout his years of study. He currently is pursuing his master's degree in information technology. His areas of interest are artificial intelligence, arts, and computational creativity.

Boshra Haghighi (originally from Iran) received her B.Sc. in systems and networking (Hons. 2009) and M. Sc. in information technology (2012) from Universiti Tenaga Nasional, Malaysia. She served as research assistant for 18 months to Azlan Iqbal under the eScienceFund research grant (01-02-03-SF0240). Her main interest is working as academic and researcher. Her research interests include human–computer interaction, e-learning, information systems, and artificial intelligence.

Chapter 1
Introduction

Abstract Computational creativity, a relatively new sub-field of artificial intelligence (AI), focuses on making machines perform tasks or produce objects that would be deemed creative by human standards. Also relevant are the methods through which this can be achieved. The field is particularly relevent today when AI has already achieved numerous milestones in tasks that were once considered doable by only humans. The most prominent example may be the defeat in 1997 of world chess champion at the time, Garry Kasparov by IBM's Deep Blue supercomputer in a 6-game match held under tournament conditions. The roots of computational creativity, however, stem not only from AI but also psychology and philosophy. Some philosophers, for example, have had issues with the very idea that computers could in any way be as creative as a human, if at all; much less that they could possibly exceed the creative capacity of a human being. Regardless, the experimental evidence emerging from computational creativity strongly suggest otherwise. In this chapter, we briefly introduce the field to readers and outline the material in the chapters that follow.

Keywords Computational creativity · Artificial intelligence · Chess · Psychology · Philosophy · Human

Computational creativity can be classified as a relatively new sub-field of artificial intelligence (AI). In principle, it focuses on the generation of creative objects such as music and visual art. These objects are created using existing, modified and in some cases even new AI approaches or techniques. These objects are usually deemed to be creative from the perspective of humans sufficiently competent in the domain; otherwise known as 'experts' (Johnson 2012). Such systems allow for some of the burden of creative content generation to be passed from humans to machines. This is because even among humans, exceptional or prolific creativity is relatively rare. Creative machines also have the potential of challenging humans with new ideas not considered, discovered or explored by humans before.

While human thought usually has many constraints due to their environment and upbringing, there are no such limitations with computers. The *way* these objects are produced may also be relevant to their creative value in some cases (Colton 2008).

A. Iqbal et al., *The Digital Synaptic Neural Substrate*,
SpringerBriefs in Cognitive Computation, DOI 10.1007/978-3-319-28079-0_1

An object of creative value might be deemed so based on its utility or beauty, or both. Remarkable products of engineering and design should therefore not be excluded by default. Notably, there is also the distinction between a 'P-creative' idea or object and an 'H-creative' one, where the former is considered something new to the *person* who generated it whereas the latter is new *historically* (Ritchie 2007; Boden 2009).

There is still something significant in terms of design, at least, about computer systems that are capable of generating P-creative content even though the ultimate goal may be H-creative works. If two people independently make a significant scientific discovery at about the same time, is their achievement any less significant than if only one of them had done so? On the other hand, if one had done so much earlier than the other, does that necessarily warrant a dismissal of the other's similar work done later but nevertheless independently? Perhaps it can be argued that the later discoverer is no less intelligent or creative than the first by virtue of being able to independently come up with the same thing. If so, then the same benefit should unbiasedly be afforded to machines.

It can be argued that chance or serendipity plays an undeniable role in creativity as well (Pease et al. 2013) and to some extent this can be simulated. In any case, we will avoid being drawn into philosophical musings about the 'true' nature of creativity or beauty (Sibley 1959; Apter 1977). Instead, we shall abide by the standard and most reasonable conventions of assessment pertaining to creativity in the particular domain of investigation and also rely upon the judgements of human experts in that domain (Didierjean and Gobet 2008). As with all computational approaches, the method must be described in sufficient detail so as to facilitate reproducibility, but due to the nature of creativity in *humans*—whom typically are not expected to explain themselves in such detail—this removes some of the 'mystery' and may affect the perception of quality with regard to the creative objects produced by the system.

Chess problems, the main domain of investigation in this book, typically requires considerable expertise, experience, time and most importantly, creativity, to produce. For this reason, and given our experience with the game in previous research works, we sought to test our approach in terms of its ability to facilitate 'creativity' in the process of composing chess problems. The game of chess has always been a favorite domain of investigation for researchers in artificial intelligence, especially, and we found that to be true here as well. The reason is that it provides a controlled and computationally-amenable environment to perform experiments and test hypotheses, as will be evident later in the book.

Success in this domain, given its complexity, therefore suggests scalability of the approach and at least *potential* in others which are beyond the scope of the present work. Chapter 2 provides a brief history and overview of some existing work in the field of computational creativity. Chapter 3 details the proposed approach and the experimental methodology we used. In Chap. 4 and its subsections we present and discuss the experimental work. Chapter 5 consolidates the experimental results in point form with a discussion about the possible limitations of the approach and how it may be applied in other domains. Chapter 6 concludes with directions for further work.

References

Apter MJ (1977) Can computers be programmed to appreciate art? Leonardo, pp 17–21
Boden MA (2009) Computer models of creativity. AI Mag 30(3):23
Colton S (2008) Creativity versus the perception of creativity in computational systems. In: AAAI spring symposium: creative intelligent systems pp 14–20
Didierjea A, Gobet F (2008) Sherlock Holmes—an expert's view of expertise. Br J Psychol 99 (1):109–125
Johnson CG (2012) The creative computer as romantic hero? computational creativity systems and creative personae. In: Proceedings of the third international conference on computational creativity, pp 57–61
Pease A, Colton S, Ramezani R, Charnley J, Reed K (2013) A discussion on serendipity in creative systems. In: Proceedings of the fourth international conference on computational creativity, p 64
Ritchie G (2007) Some empirical criteria for attributing creativity to a computer program. Mind Mach 17(1):67–99
Sibley F (1959) Aesthetic concepts. Philos Rev, pp 421–450

Chapter 2
Review

Abstract One of the first challenges researchers in artificial intelligence (AI) tried to conquer was programming a computer to play chess that could compete on the same level as humans. It took almost 50 years but once achieved, they had to find other ways of proving the value of AI. Of particular interest was the idea or concept of programmable general intelligence that the human brain possessed. Naturally, other sub-fields spawned and looked into different aspects of general intelligence that were hoped would come together synergetically. Popular culture embraced AI with its utopian and dystopian ideas of highly advanced machines and robots working for, but usually eventually against, their human creators. Expectations therefore grew and AI researchers found themselves struggling to keep up and find funding for their proposals. Computational creativity, among other sub-fields of AI such as artificial life, artificial consciousness and machine ethics even made us question what it means to be human. In this chapter, we briefly review the path AI has taken, the factors that may have influenced it and the importance of advances in computational creativity in the second millennium.

Keywords Computational creativity · Artificial intelligence · Chess · Science-fiction · Approach · Technique · Brain

Sternberg and Kaufman (2010) offer a good coverage of the concept and theories of creativity across various disciplines (primarily psychology) but these are not particularly relevant here. The *concept* of creativity with regard to artificial intelligence (AI) , however, is pertinent. It was mentioned in the proposal that lead to the famous 1956 Dartmouth Summer School often remembered as the birth of the field (Boden 2009). Even so, the field was in its infancy back then and challenges specific to creativity as opposed to the 'mere' mechanization of tasks requiring human intelligence were generally unheard of (McCorduck 2004; Levy 2006; Ekbia 2008; Warwick 2011). Put simply, getting a computer to play *good* even though bland chess was enough of a challenge then than trying to make it play good and *creatively*, e.g. in a way that would surprise even the most stylistic human grandmasters such as the late Mikhail Tal. Analogously, generating a relatively simple

A. Iqbal et al., *The Digital Synaptic Neural Substrate*,
SpringerBriefs in Cognitive Computation, DOI 10.1007/978-3-319-28079-0_2

painting was enough of a computational challenge than creating a masterpiece (Cohen 1999).

With the defeat of then-world chess champion Garry Kasparov to IBM's Deep Blue supercomputer in 1997 (Newborn 1997) and the general successes of AI in many small but significant ways in other fields (e.g. medical diagnosis, law, stock trading), demand and expectations from machines grew. Even with the resounding success of computer chess today such as computer programs running on smartphones that play at the grandmaster level and computers being indispensable in the training of human grandmasters, some experts still wonder, *where is it going?*[1] The many popular science-fiction films such as *Colossus*: *The Forbin Project* (Sargent 1970), *Blade Runner* (Scott 1982), *Terminator 2* (Cameron 1991), *The Matrix* (Wachowski and Wachowski 1999) and *A.I.* (Spielberg 2001) depicting intelligent and lifelike machines may also have had an influence on societal expectations from the field, regardless of their generally dystopian outlook.

These productions are becoming more common and are dealing with even more sophisticated issues related to AI such as human-machine social interaction in the film, *Her* (Jonze 2013; Saunders et al. 2013) and the transcendence of genuine human existence into the digital domain such as in the motion picture, *Transcendence* (Pfister 2014). Computational creativity therefore began to receive more serious attention from academia at the turn of the second millennium (Colton and Steel 1999; Buchanan 2001) perhaps to address some of these expectations and new challenges, in particular those related to the burgeoning Internet (Battelle 2006). While most applications in this area are generally small and task-specific as is also the case with traditional AI, others are far more ambitious and aim to replicate the workings of the entire human brain.

These would include projects that hope to 'give rise' to creativity and perhaps even what is known as the technological 'singularity' (Holmes 1996; Seung 2012; Kurzweil 2012). This is when AI supposedly will overtake human intelligence and radically alter the world we live in. Even an attempt to mimic the functioning of a cat's brain (Ananthanarayanan et al. 2009) was mired in controversy about its actual performance (Shachtman 2009) so we can be fairly skeptical. There are also approaches that lie somewhere between traditional AI and computational creativity such as IBM's *Watson* supercomputer which is apparently capable of extracting information and determining proper context; in English, at least. In short, it can 'read' and answer questions meaningfully (Chandrasekar 2014). However, it cannot create any novel content of its own, such as producing a book like this.

In principle, many of the fundamental approaches or techniques[2] that have been developed in AI such as artificial neural networks, genetic algorithms and evolutionary computation (McCulloch and Pitts 1943; Box 1957; Fraser and Burnell

[1]Kasparov, G. Personal Communication (with main author). 25 April 2014.

[2]This is what makes them *fundamental*, i.e. they work to a reasonable extent and have had demonstrable applications beyond a single domain or task. Note that being *task*-specific is even more constrained than domain-specific. There is also the 'no free lunch theorem' to consider.

1970; Back et al. 1997) can and have been applied to varying degrees in computationally creative systems (Abe et al. 2006; Terai and Nakagawa 2009; Correia et al. 2013; Machado and Amaro 2013). Such systems may also take the approach of combining information or knowledge from within the same domain using mathematical logic, statistical modeling or some form of machine learning, which is also used in mainstream AI (Cope 2005; Eigenfeldt and Pasquier 2013). A tempting idea is for a computationally creative system to also learn from its own 'experience', much like humans are thought to do (Iqbal 2011; Grace et al. 2013).

However, the methods these systems use tend to be domain or task-specific as well. There is nothing inherently wrong with this if the system performs well but it is certainly not any kind of *general* intelligence or 'creativity machine' that the human brain is; some brains more so than others.

In summary, the arsenal of approaches and techniques available in mainstream AI have trickled down to computational creativity and given rise to new types of applications that apply them somewhat; examples include those that generate creative objects, and assess beauty or 'interestingness' (Iqbal et al. 2012; Pérez and Ortiz 2013). At the same time, computational creativity researchers have found new ways to adapt mathematical logic and existing AI to suit their purposes. However, we could not find any approach documented in the literature that was able to successfully integrate discrete information from two or more unrelated domains so as to produce viable creative output in either one of the original sources. This is different from producing simultaneous outputs in more than one domain based on a particular set of inputs and with the help of an AI engine (Benghi and Ronchie 2013). The successful integration of discrete information from multiple, unrelated domains for computational creativity purposes may be important if one subscribes to the notion that creativity in a particular domain can be borne out of 'inspiration' obtained from more than one domain.

Take, for example, a musical piece composed that the composer says was inspired by the photo or memory of a beautiful companion. Whether or not such information in the brain is aggregated through systematicity (Phillips 2014; Clement and Gentner 1991) is presently unproven and therefore not assumed to be something our approach should be based on. So is codification of context-specific, high-level concepts really necessary for this or do low-level 'mindless' mathematical interactions suffice? The process of biological evolution is said to be a mindless and semantic-independent one that actually produces minds and even 'free will' (Dennett 1996)[3] even though there is much debate regarding the nature of the latter, particularly in association with the notion of 'consciousness' (Chalmers 1995; Harris 2012).[4] This is worth noting because consciousness and free will are often held by many as prerequisites to creativity.

The idea held by many dualists is that 'genuine' (and general) creativity cannot be described in terms of a computable, and hence fundamentally mathematical or

[3]Dennett, D.C. Personal Communication (with main author). 27 August 2012.

[4]Chalmers, D.J. Personal Communication (with main author). 17 November 2012.

algorithmic, process. In the specific case of chess problems there has never, to our knowledge, been an approach to automatic composition that relied on a 'creative process' that was independent of human involvement and that could also, in principle, be applied to other areas. Iqbal (2008, Sect. 2.4) provides a more detailed review of the main approaches in this regard that have been used in the past.

In the following chapter, we nevertheless present our novel approach to computational creativity as applied to chess problems. We have intentionally left out the numerous other techniques in AI and computational creativity to compare against to retain focus on the main subject matter of this book and more importantly, because they are fundamentally different to the one we propose even though they may be applicable to the same problem discussed. Ours, however, not only works given the aforementioned problem of quality automatic chess problem composition but is also scalable, in principle, to other domains. We therefore present our research in the practical spirit of fundamental science in that if something can be shown to work—even in a limited domain initially—*why* it works is of secondary importance. In fact, we readily concede there are open questions of this nature which are briefly discussed in Chap. 6.

References

Abe K, Sakamoto K, Nakagawa M (2006) A computational model of metaphor generation process. In: Proceedings of the 28th annual meeting of the cognitive science society, pp 937–942

Ananthanarayanan R, Esser SK, Simon HD, Modha DS (2009) The cat is out of the bag: cortical simulations with 109 neurons, 1013 synapses. In: High performance computing networking, storage and analysis, proceedings of the conference on IEEE, pp 1–12

Back T, Hammel U, Schwefel HP (1997) Evolutionary computation: comments on the history and current state. Evol Comput IEEE Transac on 1(1):3–17

Battelle J (2006) The search: how google and its rivals rewrote the rules of business and transformed our culture. Penguin Group (USA) Inc, Portfolio

Benghi C, Ronchie G (2013) An artificial intelligence system to mediate the creation of sound and light environments. In: Proceedings of the fourth international conference on computational creativity, p 220

Boden MA (2009) Computer models of creativity. AI Magazine 30(3):23

Box GE (1957) Evolutionary operation: a method for increasing industrial productivity. Appl Stat, pp 81–101

Buchanan BG (2001) Creativity at the metalevel: AAAI-2000 presidential address. AI Magazine 22(3):13

Cameron J (Director) (1991) Terminator 2: judgment day [motion picture]. Carolco Pictures, United States

Chalmers DJ (1995) Facing up to the problem of consciousness. J Conscious Stud 2(3):200–219

Chandrasekar R (2014) Elementary? question answering, ibm's watson, and the jeopardy! challenge. Resonance 19(3):222–241

Clement CA, Gentner D (1991) Systematicity as a selection constraint in analogical mapping. Cogn Sci 15(1):89–132

Cohen H (1999) Colouring without seeing: a problem in machine creativity. AISB Quarterly 102:26–35

Colton S, Steel G (1999) Artificial intelligence and scientific creativity. Artif Intell Study Behav Q 102
Cope D (2005) Computer models of musical creativity. MIT Press, Cambridge
Correia J, Machado P, Romero J, Carballal A (2013) Evolving figurative images using expression-based evolutionary art. In: Proceedings of the fourth international conference on computational creativity, p 24
Dennett DC (1996) Darwin's dangerous idea: evolution and the meanings of life. Simon and Schusterm, New York
Eigenfeldt A, Pasquier P (2013) Considering vertical and horizontal context in corpus-based generative electronic dance music. In: Proceedings of the fourth international conference on computational creativity, p 72
Ekbia HR (2008) Artificial dreams: the quest for non-biological intelligence. Cambridge University Press, Cambridge
Fraser A, Burnell D (1970) Computer models in genetics. McGraw-Hill, New York
Grace K, Gero J, Saunders R (2013) Learning how to reinterpret creative problems. In: Proceedings of the fourth international conference on computational creativity, p 113
Harris S (2012) Free will. Simon and Schuster, New York
Holmes B (1996) The creativity machine. New Scientist, pp 22–26
Iqbal MAM (2008) A discrete computational aesthetics model for a zero-sum perfect information game, Ph.D. thesis. University of Malaya, Kuala Lumpur, Malaysia. https://www.researchgate.net/publication/230855649_A_Discrete_Computational_Aesthetics_Model_for_A_Zero-Sum_Perfect_Information_Game
Iqbal A (2011) Increasing efficiency and quality in the automatic composition of three-move mate problems. In: Anacleto J, Fels S, Graham N, Kapralos B, Saif El-Nasr M, Stanley K (eds) Entertainment computing—ICEC 2011, lecture notes in computer science, vol 6972, pp 186–197. 1st edition, 2011, XVI. Springer, Berlin. ISBN 978-3-642-24499-5
Iqbal A, van der Heijden H, Guid M, Makhmali A (2012) Evaluating the aesthetics of endgame studies: a computational model of human aesthetic perception. IEEE Trans Comput Intell Al Games: Spec Issue Comput Aesthetics Games. 4(3):178–191. ISSN 1943-068X. e-ISSN 1943-0698
Jonze S (Director) (2013) Her [motion picture]. Annapurna Pictures, United States
Kurzweil R (2012) How to create a mind: the secret of human thought revealed. Penguin, New York
Levy DN (2006) Robots unlimited: life in a virtual age. AK Peters, Wellesley
Machado P, Amaro H (2013) Fitness functions for ant colony paintings. In: Proceedings of the fourth international conference on computational creativity, p 90
McCorduck P (2004) Machines who think: a personal inquiry into the history and prospects of artificial intelligence. AK Peters Ltd, Natick
McCulloch WS, Pitts W (1943) A logical calculus of the ideas immanent in nervous activity. Bull Math Biophys 5(4):115–133
Newborn M (1997) Kasparov vs. Deep Blue: Computer Chess Comes of Age. Springer, New York
Pérez RPY, Ortiz O (2013) A model for evaluating interestingness in a computer-generated plot. In: Proceedings of the fourth international conference on computational creativity, p 131
Pfister W (Director) (2014) Transcendence [motion picture]. Alcon Entertainment, United States
Phillips S (2014) Analogy, cognitive architecture and universal construction: a tale of two systematicities. PLoS ONE 9(2):e89152
Sargent J (Director) (1970) Colossus: the forbin project [motion picture]. Universal Pictures, United States
Saunders R, Chee E, Gemeinboeck P (2013) Evaluating human-robot interaction with embodied creative systems. In: Proceedings of the fourth international conference on computational creativity, pp 205–209
Scott R (Director) (1982) Blade runner [motion picture]. Warner Bros, United States
Seung S (2012) Connectome: how the brain's wiring makes us who we are. Houghton Mifflin Harcourt, United States
Shachtman N (2009) Darpa's simulated cat brain project a 'scam': top scientist. Wired Magazine. http://www.wired.com/2009/11/darpas-simulated-cat-brain-project-a-scam-top-neuroscientist/

Spielberg S (Director) (2001) A.I. [motion picture]. Warner Bros, United States
Sternberg RJ, Kaufman JC (eds) (2010) The Cambridge handbook of creativity. Cambridge University Press, Cambridge
Terai A, Nakagawa M (2009) A neural network model of metaphor generation with dynamic interaction. In: Alippi C, Polycarpou M, Panayiotou C, Ellinas G (eds) ICANN 2009, Part I. LNCS, 5768. Springer, Heidelberg, pp 779–788
Wachowski A, Wachowski L (Directors) (1999) The matrix [motion picture]. Warner Bros, United States
Warwick K (2011) Artificial intelligence: the basics. Routledge, London

Chapter 3
Methodology

Abstract We explain in detail the workings of our proposed computational creativity approach, i.e. the Digital Synaptic Neural Substrate (DSNS). It is a new computational technique that is able to combine fragments of information from different domains such as music and paintings in order to create new objects in any of them. The DSNS is therefore in a sense much like the human brain which stores perceived information from its surroundings and occasionally produces creative output such as music, paintings and chess problems. This approach serves as the 'spark' of creativity and the supplier of specifications in the creation of objects deemed potentially to have creative value. In principle, there is no limit to the number of domains, objects or output that can be created. This makes the DSNS highly scaleable and adaptable; analogous to how artificial neural networks and genetic algorithms can be applied in many areas and for many purposes.

Keywords Computational creativity · Intelligence · DSNS · Approach · Technique · Brain · Music · Painting · Chess

3.1 Overview

The Digital Synaptic Neural Substrate (DSNS) approach[1] was intended to provide artificial intelligence (AI) researchers with a generic method to combine information, or rather data, from different domains (e.g. chess, music, paintings) such that they could be integrated in a way that led to the automatic creation of new, original objects from any one of those domains. This can be thought of as a kind of catalyst to the 'spark' of creativity, if you will. A useful analogy may be the creation of a new painting by a human after having been exposed to or 'inspired' by a musical

[1] Iqbal (2014).

A. Iqbal et al., *The Digital Synaptic Neural Substrate*,
SpringerBriefs in Cognitive Computation, DOI 10.1007/978-3-319-28079-0_3

piece or a different painting *and* a musical piece. It would be difficult for the human painter to say for certain which aspects of each led to which aspects of the new painting (or how) except that these two objects 'come to mind' when asked about the 'inspiration' behind the new painting.

It could also be that only one such object is mentioned while the other(s) remain buried deep in memory or the subconscious. We could not find any significant and descriptive neuroscientific or neurological bases for the formation of creative thought so researchers in other areas should be free to posit their own ideas for testing. There is also no 'requirement' for a functionally creative computational process to be grounded in the method employed by the *human* brain, whatever it may be. Typically, it should be self-evident that the human mind or rather brain must contain bits and pieces of information from various objects we have perceived through our experiences in the form of generic neuro-chemical substances (the 'format' of the brain) and is able to integrate them through volition in poorly-understood ways in order to (sometimes) create new objects of creative value.

At present, this is experimentally somewhat beyond human efforts considering, for example, fMRI and brain hemodynamics which simply identify regions of the brain related to particular tasks or activities (Ruiz et al. 2014; Strang et al. 2014). The aforementioned basis of the formation of creative thought is therefore not mere conjecture. It is virtually unheard of that a person should be creative in mathematics knowing absolutely nothing about it. Some basic pieces of knowledge about mathematics—along with pieces of all sorts of other information the person has perceived and retained—must first reside in the brain. The precise method through which these pieces of information are intermingled (in particular with regard to producing creative objects or ideas) is far from understood so we freely posit our own idea to be tested. With this concept of creative information processing in mind, the 'DSNS' terminology and approach was developed. It works as follows.

3.2 The DSNS Approach Explained

First, we need two collections, sets or samples of objects from the same or different domains. For example, 100 chess problems and 100 paintings (two domains); or 200 of either divided into two subsamples (one domain). Each object, depending on the domain it is from, is described using a set of attributes. A chess problem may be described using the attributes: *the number of white pieces on the board*, *the Shannon value of those white pieces* and *the difference in value between the white and black pieces*. These are things that could be perceived or found out about the objects in question. There is theoretically no limit to the number of attributes that can be used to describe an object. However, it should be numerically representable. A painting, on the other hand, may be described using the attributes: *the number of pixels in the image*, *the number of colors used* and *the year it was painted.*

It is acknowledged that these attributes may not be related in any meaningful way. This is intentional to reflect the influx of various types of conscious and subconscious information into the brain. Put simply, if two people were to observe the same object for a given period of time, one person might remember a subset of its attributes and the other a different, though probably somewhat overlapping, subset of attributes. These attributes are the things that would help these people describe the object they saw and perhaps even recreate something like it. They may also be able to incorporate these attributes into a new object of the same class but original in its own right. In selecting attributes, one might prefer unique attributes in the domain that do not 'overlap' much. For example, *the number of pixels in the image* should not be included along with say, *the number of 16 × 16 blocks of pixels in the image* because these are simply variations of the same concept. *The number of white pieces on the board* should not be included along with *the number of white pawns on the board* because the latter is a subset of the former.

There are no fixed rules with regard to attribute selection (which is perhaps a strength when it comes to creativity) and we hesitate to state any but a little common sense and good judgement can go a long way. The number of attributes in a domain therefore also becomes more manageable. It is presumed that we have at least some basic knowledge about a domain sufficient for us to identify at least some of the attributes that may partially describe it. So it is fair to say that if we have absolutely no knowledge about a domain, we cannot identify and represent any of its attributes, much less in the form of numbers.

Second, the attribute values for each object should be tabled as a single row along with its object identifier. This row and its values can be identified as a DSNS *string*. An example of three chess problem DSNS strings is provided in Table 3.1. The Forsyth-Edwards Notation (FEN) is a representation of the starting position of the chess problem. Mainly, it describes where all the pieces are and whose turn it is. As an identifier or 'key' for the object, the FEN is typically included along with the attributes and their values to make up the DSNS string for a chess problem. So in Table 3.1, each row, including the FEN, is a separate DSNS string.

An example of DSNS strings for three paintings is shown in Table 3.2. Here, a unique number serves as the identifier for each painting. This number may simply be derived from the paintings as they are arranged alphabetically based on their

Table 3.1 Example DSNS strings for three chess problems

FEN	White pieces	Black pieces	Value of white pieces	Value of black pieces	Value difference
8/1p2BN1K/4Qp2/n1R4p/3k2P1/P5n1/4P3/1r6 w—0 1	8	7	23	14	9
5rk1/5qpn/8/3N4/3B4/1B6/1KP3R1/8 w—0 1	6	5	15	18	3
5Q2/b2k1P2/1n1NNn2/1P1p4/6P1/8/8/7K w—0 1	7	5	18	10	8

Table 3.2 Examples DSNS strings for three paintings

Painting ID	Pixel count	Colors	Identifiable objects	Year	Aspect ratio
115	398,321	81,935	4	1626	0.729
16	282,576	68,624	3	1513	0.762
175	430,407	41,510	2	1917	0.611

names in a list. So with 300 paintings, the *Mona Lisa* may be numbered 150 and *The Scream* numbered 250. The quantity of attributes for an object, especially between different domains, need not be fixed. In the example case presented, the chess problem sample and the painting sample each contain three individual objects with five attributes each. However, the DSNS approach also allows for an unequal number of attributes between the two domains being used.

Third, a random DSNS string is selected from each sample in order to generate a 'deviation' value. The idea behind the deviation value is that it can be thought to act as a measure of 'creative difference' between two objects. One might think of this concept as follows. The deviation value between say, a musical piece by Mozart and a similar one by Salieri would likely be smaller than the deviation value between the one by Mozart and a song by Lady Gaga. It should be noted the deviation value is not intended to act as a kind of classification system. Rather, it tries to capture somewhat the different creative *natures* of the objects being contrasted. One may just as well be a fan of Mozart *and* Lady Gaga.

Table 3.3 shows two DSNS strings (from the same domain, for illustrative purposes) and their attributes. The two columns to the right, i.e. $|d|$ and $(\Sigma\div)$ represent 'absolute difference' and 'summative division', respectively. The former is simply the absolute difference between the two values of a particular attribute given the two strings whereas the latter is a new concept; it represents the sum of the division of the first string's attribute value with the second string's attribute value, and vice versa; e.g. for attribute 1 in Table 3.3, it is 6/7 + 7/6. What we call 'summative division' was used because we wanted something simple yet functional enough to create sufficient variation in the deviation values that was also not, to our knowledge, previously described in the mathematical literature.

The deviation for the two DSNS strings is therefore = $[\Sigma|d| + \Sigma\,(\Sigma\div)] - \text{CID}$. The CID or 'creative indifference value' ('6' in this case) is obtained from the sum of the absolute differences and summative divisions (i.e. $[\Sigma|d| + \Sigma\,(\Sigma\div)]$) for two DSNS strings with *exactly the same attribute values*; a sort of baseline. In theory,

Table 3.3 Absolute difference, summative division and deviation

Attribute	String 1	String 2	$\|d\|$	$(\Sigma\div)$	Deviation
1	6	7	1	2.024	(7 + 6.275) − 6 = 7.275
2	5	7	2	2.114	
3	13	9	4	2.137	
			7	6.275	

since the attribute values are exactly the same, the objects should also be (creatively) the same. Imagine if, in Table 3.3, for attribute 1, strings 1 and 2 both had a value of 6, the absolute difference would be '0' and the summative division would be (6/6 + 6/6 = 2). This would be true for all the attributes in the strings. The sum of 0 + 2 = 2; and the number of attributes, *n*, would lead to a CID of 2*n*.

Fourth, this deviation of 7.275 (see Table 3.3) is then used to generate two *new* DSNS strings by randomly selecting possible attribute values from the original sample (so the values are 'realistic') until a pair is found that matches that deviation (i.e. 7.275) using the same process just described above.[2] So suppose the sample has 300 DSNS strings,[3] there are 300 possible values for each attribute, or somewhat less if the values for each attribute are not unique. A random change of all these attribute values, in new combinations, might lead to two *new* DSNS strings such that the same, precise deviation value results.[4] This can be called 'stage 1' of the search for a perfect match. The idea put forth here is to obtain two new strings which have a similar contrast (i.e. difference) in 'creative value' as the original pair.

If after 30 s (the amount of time is flexible but should be reasonably long), no exact match can be found using the 'realistic' attribute values in the sample, then the search is 'widened' to include *all* possible values (i.e. to one decimal place or more, depending on the nature of the attribute values and time constraints) between the lowest and highest attribute values in the sample. This is 'stage 2'. So if for a particular attribute, there were only five values in the sample: 3, 7, 9, 10, 12, now *any* value between 3 and 12 (i.e. including 4, 5, 6, 8 and 11) would be sought after. This goes on for another 30 s. If still no match is found, then the final 'stage 3' takes effect where the two new DSNS stings that produced the closest deviation to the one desired are taken. 'Contemplation time' is therefore considered a factor in the creative process.

So in the case of the 7.275 deviation, after 60 s, taking into account all the strings generated in stages 1 and 2 (which means they should be stored in an historical array of some sort), if the closest deviation produced was 7.15, then those

---

[2]Note that the deviation value can be rounded to one or two decimal places to improve performance. The requirement of a pair of objects is fundamental and built into the DSNS approach. A 'randomization' approach on the attribute values of just one DSNS string would, in principle, not provide any basis of comparison with another creative object that would have its own DSNS string. There are likely hundreds of different ways the DSNS approach could be executed differently but we had to decide on the one we expected to work best.

[3]The sample of 300, in this case, is used simply because it was the source of the original two DSNS strings and readily available. A slightly larger or even smaller 'knowledge base' (e.g. 290, 310) to look through can also be used at this point but it would likely not have a significant impact on the result. On the other hand, a much larger knowledge base of say, 600 or even 1000 objects (even from an altogether different sample than the two strings came from) might yield better results.

[4]The approach we suggest is to randomly select new values for *all* the attributes in each tentatively new string at the same time before testing to see if the new strings produce the desired deviation. Systematically changing the value of just one attribute in either string to see if it brings you closer to the desired deviation is also possible, as are presumably other methods.

two DSNS strings that produced it would be retained. These two new strings can then be used to generate one or more new objects based on the same set of attributes (but with potentially different values) used to describe the other objects. Stage 3 is necessary to prevent the system from running into dead ends. While chancing upon the precise deviation value is desirable, something close enough should suffice. The generation tool or technique is a separate issue beyond the DSNS and domain specific; in the case of *this* research, the chess problem generation process is explained in Chap. 4 and Appendix A.

In summary, the first two DSNS strings selected represent two unique objects within the domain (e.g. two chess problems). Their ‘creative difference’ is represented by the deviation value generated. This deviation value is then used to produce two new strings (which themselves would generate the same or similar deviation value) but with different *attribute* values. These new attribute values can ultimately be used to generate new objects of potentially creative value using whatever generation system is available, i.e. by feeding it the attribute values it needs to build objects of a particular type. In very simple terms, it is like having or being able to derive new (realistic) measurements of height, width and length to build a new box. The box-building system is not part of the DSNS approach itself.

3.3 Using Different Domains

If objects from two different domains should be integrated using the DSNS approach, the process is the same except that *two* deviations are generated. Each deviation is obtained by choosing two random DSNS strings from the same domain, following the steps described in the previous section. These two deviations are then ‘merged’. So if the first domain was chess problems and the second music, and the first deviation obtained was 126.21 whereas the second was 35,722.11 (the numbers can be quite different), the two deviations can be merged as shown in Fig. 3.1. If an object from the chess problem domain was desired, then the deviation value would need to relate to that domain, i.e. be the sort of number that a deviation derived entirely from objects in that domain would look like. However, with two deviations and only one of them relating to the chess problem domain, they need to be merged as shown on the left side of Fig. 3.1. This may be seen as a sort of ‘lateral thinking’ approach to merging the two unrelated deviation values so they become functionally one in the desired or target domain.

What happens is the chess problem deviation (i.e. 126.21) is placed in the first row with the music deviation (i.e. 35,722.11) directly beneath it, aligning the digits so they match. The ‘3’ and ‘5’ are simply crossed out because there are no digits of that value in the row above. The remaining digits follow this rule: *If [top digit] modulo [bottom digit] = 0 then result = [bottom digit] else result = [top digit]*. So the result here is 122.11. On the right side of Fig. 3.1, a deviation relating to the music domain is desired so that deviation is put in the first row with the chess problem domain deviation in the second. The ‘3’ and ‘5’ in this case simply ‘fall

Chess Problem Domain Desired

		1	2	6	.	2	1
~~3~~	~~5~~	7	2	2	.	1	1
		1	2	2	.	1	1

Music Domain Desired

3	5	7	2	2	.	1	1
		1	2	6	.	2	1
3	5	1	2	2	.	1	1

Fig. 3.1 Merging deviations from different domains

down' and the remaining digits follow the same rule mentioned above. The result in both cases (i.e. the last row) is a *single* deviation value that can be used as described in Sect. 3.2 (the fourth step) to generate two new DSNS strings. In some cases there may be no difference at all between the merged deviation and the original deviation from the same domain.

This process of merging may reflect or mimic the imperfect recollection of information by humans which inherently produces variation that can be functional in the creative thought process. In other words, they recall information that is similar and functional in the domain but not necessarily identical to what they have seen before. We consider this to be self-evident as well. By the same token, *more* than two domains can, in principle, be merged. The chess problem and music domain can be merged to produce a deviation of say, the chess problem type. This deviation can then be merged with a deviation from the painting domain to produce a deviation of the painting type. So data from the chess problem, music and painting domains can be combined to produce content in the painting domain. Theoretically, there is therefore no limit to the number of domains that can be merged in this way. However, this is beyond the scope of the present work.

3.4 Using Strings of Different Length

In some cases, a sample of DSNS strings may contain certain objects or 'individuals' that lack a particular attribute value. For instance, the 'year' attribute may not have been available for a particular object such as a painting. In such cases, it is still usable as a DSNS string along with other strings that actually have the year attribute value. The one that does not is simply given the value 'NULL'. During the deviation-generation process (e.g. a string with 5 attributes and one with 4 attributes +1 NULL attribute happens to be randomly selected) the one with 5 is reduced to 4 attributes by removing the corresponding NULL one. So we are left with an even 4 to 4 DSNS string length in that particular case. This is different from having an attribute value of '0', which is used like any other value. In general, a DSNS string of a different length implies that it either has fewer or more attributes.

This is typically an issue only for strings from the same domain because as explained in Sect. 3.3, using different domains implies generating two deviations (four DSNS strings involved) and only then merging the deviations. It would not be advisable trying to process two DSNS strings from the same domain but with entirely different attribute types. They should match so that the attribute values carry *some* meaning, even if occasionally the pair needs to default to using one attribute fewer due to a NULL entry in either one of them.

3.5 Assumptions and Implications

This DSNS approach is assumed to be able to generate attributes that can then be used to generate creative content. All or some of the attributes may be used for this purpose, depending on the domain and external generation system available. Essentially, we randomly select two objects from a domain, 'reduce' them to DSNS strings and then use those strings to generate a deviation value that represents the 'creative difference' between those two objects. From this deviation value, two *new* DSNS strings are generated (with a similar 'creative difference'). Each or even both of these can then be used to build or create one or more new objects that potentially have creative value. How these objects are created is a separate issue and dependent upon the domain in question and the generation or 'building' system available. Figure 3.2 shows a general diagram of the DSNS approach.

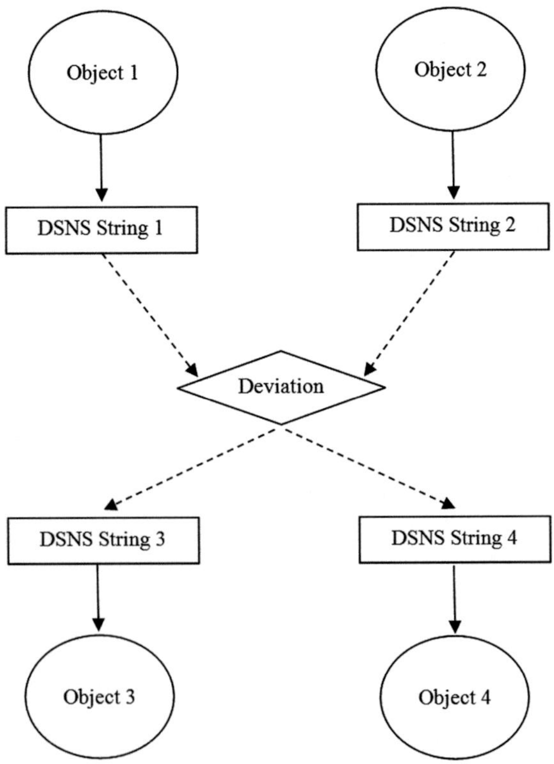

Fig. 3.2 General diagram of the DSNS approach

It is therefore implied that raw data 'extracted' from an object—even from different domains—can be integrated such that new objects of creative value from either of the original domains, can be generated. This is comparable to how a human being perceives, say, a painting, which is likely represented in memory as a neuro-chemical substance or reaction in the brain, and then a piece of music, which is also represented in the 'similar format' neuro-chemical substance or reaction in the brain (analogous to the DSNS strings) so that they can be integrated (analogous to the deviation value) to produce either a new painting or new piece of music (analogous to the building specifications supplied by the *new* DSNS strings). It stands to reason that if the deviation value (i.e. the 'creative difference') between two DSNS strings derived from two objects can be matched by two new strings with realistic attribute values other than what was contained in the original DSNS strings, then these two new strings—in particular the attribute values they contain—can be used to generate or create new objects that would resemble the objects from which the original DSNS strings were derived, i.e. in terms of creative value.

The *contrast* between the two DSNS strings which is represented by the deviation value is not necessarily a measure of how weak and strong, creatively, the two objects in question are. It could just as well be a measure of how different two equally strong (or equally weak) objects are in terms of creativity. Chances are, however, one is better or more interesting than the other, however slightly, even. This is also true for the two new strings from which two or more new objects can potentially be created.

References

Iqbal MAM (2014) A process of integrating information from different domains for the purpose of generating novel creative objects. Malaysia Patent Application No.: PI 2014703983. Filing Date: 24 Dec 2014

Ruiz S, Buyukturkoglu K, Rana M, Birbaumer N, Sitaram R (2014) Real-time fMRI brain computer interfaces: self-regulation of single brain regions to networks. Biol Psychol 95:4–20

Strang S, Utikal V, Fischbacher U, Weber B, Falk A (2014) Neural Correlates of receiving an apology and active forgiveness: an fMRI study. PLoS One 9(2):e87654

Chapter 4
Experimental Work

Abstract We demonstrate how the Digital Synaptic Neural Substrate (DSNS) approach can be applied in the domain of chess problem composition (an area requiring creativity) using fragments of information from classical music, renowned paintings, photos, chess games and other chess problems. Due to the sheer volume produced by computer, the quality of the compositions is assessed using an existing and experimentally-verified computational chess aesthetics model incorporated into a computer program called Chesthetica. In addition, human expert consultation is also used. The experimental results suggest that higher quality compositions can be generated by combining fragments of information from photographs of people and chess games between weak players. While the reasons for this and our other findings remain open questions for now, directions for further work, perhaps even in different domains, are clear. The experimental setups presented should also prove useful to other researchers (not just in artificial intelligence) hoping to replicate or expand upon the findings.

Keywords Creativity · Aesthetics · Chess · DSNS · Chesthetica · Music · Painting · Photo · People · Expert

4.1 The Domain of Investigation

In order to test if the DSNS approach actually worked, we had to first select a domain of investigation that required creativity (Hesse 2011). Given our prior experience and background, the domain of chess problems or compositions was selected. In particular, three-move, forced-mate problems (i.e. against any defense) otherwise known as 'three-movers' (#3, for short) since it is one of the most common and illustrative. Chess problems are basically like puzzles with a stipulation (e.g. *White to play and checkmate in three moves*) and the solver is challenged to discover the winning sequence of moves. In most cases, the solution is somewhat unexpected which is why problems are often considered works of art and beautiful or aesthetic in

A. Iqbal et al., *The Digital Synaptic Neural Substrate*,
SpringerBriefs in Cognitive Computation, DOI 10.1007/978-3-319-28079-0_4

nature (Humble 1993, 1995; Lord 1984–85; Rachels 1984–85; Ravilious 1994). Not all 'working' or valid problems, however, are equally aesthetic. For instance, the final forced three-move mate sequence taken from a tournament game between two expert players might *resemble* a composed chess problem in terms of the basic requirements but it lacks the forethought and 'design' that a human composer puts into a chess problem intended for publication in a magazine or book.

Figure 4.1 a, b shows a typical three-mover and three-move checkmate sequence taken from an actual game between two expert players, respectively. The two are 'physically' similar in the sense that we have a starting position (in the case of the real game it is taken just prior to move 37) and the existence of a forced mate-in-three sequence. They differ in the sense that the composition (a) was designed to be economical and attractive whereas the real game occurred over the board in a tournament. Composition convention dictates that compositions should be designed such that they *could* have occurred in a real game (e.g. you cannot have a black bishop on the **a8** square *and* a black pawn on the **b7** square) and *should* appear realistic (e.g. you should not have 6 queens and 4 rooks of the same color).

Compositions also tend to feature chess themes or motifs such as the *grimshaw* or *plachutta* with varying degrees of sophistication in their solutions, i.e. the main line of play and possible variations (Velimirovic and Valtonen 2013). These can be planned during the design phase of a chess problem or they could simply emerge in a tournament game winning sequence. A good (as in of publishable quality) composition may therefore take an expert human composer hours or even weeks to compose; perhaps even longer. A grandmaster (GM) title in composing as awarded by the World Chess Federation (FIDE), for example, can take decades of experience, work and achievements to obtain. John Rice, a leading British composer who started composing at age 14, only recently obtained the GM title at age 78 (Nunn 2015). For

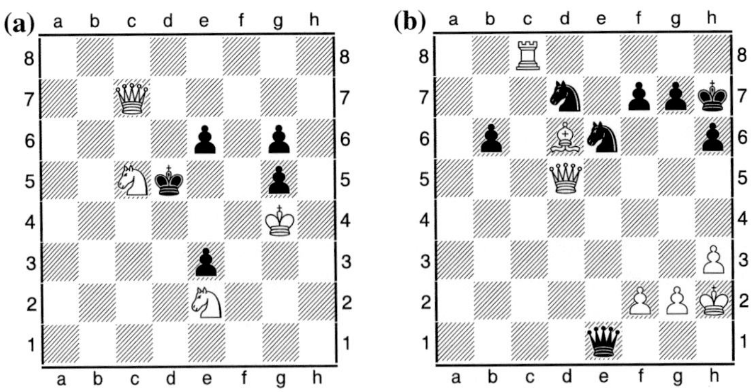

Fig. 4.1 A typical #3 chess problem and forced three-move mate sequence from a real game. **a** Vincente Maria Norberto Portil The Westminster Papers 1872, 1. Ne4 Kxe4 2. Qd8 e5 3. Ng3# 1-0. **b** Wells, P. K. versus Baburin, A. BCF-chT 9900 (4NCL) 1999, 37. Qf5+ g6 38. Qxf7+ Ng7 39. Qg8# 1-0

these reasons, compositions are considered, on average, to be more attractive or aesthetic than sequences taken from real games (even between experts) where 'design' factors cannot be controlled and the players are furthermore typically under time constraints to win by any legal means possible, rather than to win artfully. Some of these issues are explored in greater detail in (Iqbal 2008), for the interested reader.

4.2 The Experimental Tool

We used an experimentally-validated computational chess aesthetics model to evaluate the chess problems and tournament game sequences mentioned in the previous section. Good references with regard to the field of computational aesthetics can be found in (Iqbal 2015; Galanter 2012). This model was incorporated into a computer program called CHESTHETICA (Iqbal et al. 2012). Generally, the model uses formalizations of a selection of aesthetic principles and themes derived from the chess literature, in conjunction with stochastic technology (i.e. some randomness) to evaluate beauty in #3 sequences. We found it advisable to use the formalization or 'equations' approach to aesthetics[1] even though other approaches (e.g. an artificial neural network) may be effective as well; at present, however, these remain untested. Assessments based on the model—typically in the form a single digit number precise to three decimal places—correlate positively and well with domain-competent human aesthetic evaluation (Iqbal et al. 2012).

A notable characteristic of the program is that it may produce a slightly different aesthetics score the second or third time it evaluates the same move sequence. However, this is generally inconsequential so only one cycle of evaluation was used per chess problem, i.e. not an average of 10 or 20 cycles per problem which would have been a waste of time and resources. We have not found a way to apply the model in its entirety as an heuristic which can be used in any sort of known hill-climbing search because the dynamics of the chessboard are not really predictable; which is what makes it such a creative endeavor and full of paradoxes. As an analogy, the ability to assess the beauty of an object (what the model does) does not automatically render that ability capable of producing such objects or improving upon them continuously in a systematic way. We will not go into further detail about the model here or try to revalidate it in contrast to other generic ideas on aesthetics (Ritchie 2007). Interested or skeptical readers are invited to review the aforementioned reference for a complete explanation.

A downloadable version of CHESTHETICA ENDGAME (CEG) which supports both #3 and endgame studies is also available (Iqbal 2012). Endgame studies are typically a longer, more sophisticated type of chess problem.[2] The reason we used this computational aesthetics model to perform the aesthetic evaluations of the

[1]McCarthy, J. Personal Communication (with main author). 17 September 2005.

[2]Nunn, J. Personal Communication (with main author). 22 February 2011.

three-movers is simply because it would be too much to ask of human experts to assess hundreds of problems and sequences. Doing so would also likely result in less consistent results due to fatigue and human error. CHESTHETICA v9.22 was modified and also used to automatically compose three-movers based on the DSNS approach. It is therefore the 'generation tool' in this particular case that relies on the new DSNS string attribute values, as alluded to in Chap. 3, Sect. 3.2. The newer composing module is independent of the program's aesthetics evaluation components mentioned earlier. The program's earlier composing module included a 'random' and 'experience table' approach not particularly relevant to this research (Iqbal 2011). Details about how the existing composing algorithm was modified to accommodate the DSNS approach is provided in Appendix A.

The program is capable of imposing a 'filter' of up to 5 common composition conventions, namely, *no cooked problems*, *no duals*, *no checks in the key (i.e. first) move*, *no captures in the key move* and *no key move that restricts the enemy king's movement*. Only the third one was applied for all experiments to keep the composing rate reasonably high and consistent for experimental purposes. The composing rate of CHESTHETICA drops exponentially the more filters are applied. This means it would take very much longer to get the same number of compositions. In general, a composition adhering to only one convention is considered of lower quality than one adhering to say, five. However, this is not necessarily true in every case and there is some evidence to suggest that composition conventions may not be as important to non-esoteric aesthetics as composers tend to think they are (Iqbal 2014). It is also arguable that in attempting to abide by more and more conventions in the composing process, the 'paths' toward higher quality compositions may be blocked or never become available. Say, for instance, a composer imposes a convention upon himself that only the original set of pieces (i.e. no promoted pieces) may be used; the possibility of a very interesting or attractive composition with three or four knights of the same colour would then be completely ruled out.

4.3 The First Experiment: Same Domain

The first experiment was intended to test the hypothesis that the DSNS approach—using one domain as a source—can be used to generate objects of creative value in the same domain, i.e. forced three-movers (compositions and from tournament games). The term 'creative value' here implies *aesthetic* value. Aesthetics was chosen as a measure of creativity because it stands to reason that beautiful man-made objects are typically the product of creative minds. We also intended to use the experimental tool, CHESTHETICA (see previous section) which is the only known experimentally-validated chess aesthetics evaluation software in the world. The Meson Chess Problem Database (http://www.bstephen.me.uk) was used as a source of compositions by (largely) experienced human composers. From a total of 29,453 three-movers, 300 assessed to have an aesthetics score above 3.5

(by CHESTHETICA) were used.[3] Another 300 assessed to have a score below 1.25 were also used. The two sets represent both high and low quality samples in terms of creative content. Each of these sets were randomly divided into two subsets of 150 compositions each, for a total of four sets or subsamples.

Forced three-move checkmate sequences taken from tournament games were also used. These were taken from the ChessBase Big Database 2011. From a total of 5,063,778 games, 300 games between players with an Elo rating above 2500 were used. Another 300 between players with an Elo rating below 1500 were used. These two sets also represent high and low creative content, respectively. Each of these sets were further divided into two subsets of 150 compositions each, for a total of four sets or subsamples. It stands to reason that with higher quality source materials, the DSNS approach should produce chess compositions of higher quality, on average, than with lower quality materials. For the compositions, the two subsets of 150 compositions were used to generate the deviation values (see Sect. 3.1) and produce new pairs of DSNS strings for use by CHESTHETICA in composing new chess problems (see Appendix A). The same was done for the tournament game subsets. For brevity, these sets will be referred to as Comp3.5, Comp1.25, TG2500 and TG1500.

The ten attributes used in the DSNS process for this domain represent some feature of the chess problem or tournament game sequence that a human observer might notice or be able to find out. They are, in principle, arbitrary, but should be obtainable by some means and describable using real numbers. These include:

1. The number of white pieces in the initial position
2. The number of black pieces in the initial position
3. The Shannon value of those white pieces
4. The Shannon value of those black pieces
5. The difference between the two Shannon values (i.e. material difference)
6. The number of moves in the sequence (fixed at 3)
7. The year the chess composition was composed/tournament game was played
8. The first piece to move in the sequence (P = 1, N = 2, B = 3, R = 4, Q = 5, K = 6)
9. The last piece to move in the sequence (P = 1, N = 2, B = 3, R = 4, Q = 5, K = 6)
10. The sparsity value of the initial position

The idea here is that these 'scraps' of information are similar to how the brain stores pieces of information from objects we observe. It is clear that few, if any, of us can remember objects in precise detail. More likely is that we remember particular attributes of an object that appeal or have some significance to us. Taken collectively, these features represent that object in our brains. The first two

[3]Aesthetics scores by the program typically range between 0 and 5 with hardly any at either extreme (none known to date, in fact). It is also difficult to determine what the highest possible aesthetics score for a three-move sequence could be given the dynamics and virtually infinite possibilities on the chessboard.

attributes are clear. The third is the total value of the chess pieces as described by Claude Shannon in his seminal paper on computer chess (Shannon 1950). He attributed relative values to the chess pieces as follows: queen = 9, rook = 5, bishop/knight = 3 and pawn = 1. Incidentally, Alan Turing proposed slightly different values, i.e. P = 1, N = 3, B = 3.5, R = 5, Q = 10 (Turing 1953). The king is of infinite value because losing it means losing the game. With approximate values such as this, computers are able to gain a good idea of the material imbalances on the board and play a decent game of chess. Modern chess programs may use slightly different weights and hundreds of other game-playing heuristics but these values still serve well as a rough guide with regard to which army is doing better; sufficient for the purposes of our experiments. The fifth attribute is clear.

The sixth attribute is the same for all the chess problems, i.e. '3'. In future implementations of the DSNS where chess problems of different lengths may be used, this attribute will vary, of course. The fact that this one attribute has the same value for all compositions and sequences actually makes the DSNS experiment more realistic because it is unlikely that objects from any domain would have attributes that differed in *every* regard. Paintings may have the same frame size or weight whereas musical tracks may all have the same frequency. All the attributes values taken collectively, however (as is the case in the DSNS approach) still provide for sufficient variation in the new DSNS strings to be generated. This is also not equivalent to 'artificially' leaving the sixth attribute out and using only nine. The seventh attribute is more subtle and not obvious from the moves itself. The sequence could have been composed or played yesterday or 500 years ago. This piece of information is nevertheless something that a user may care to find out by checking the game details available in most chess databases or by doing an online search.

The eighth and ninth attributes are integer codes that represent the first and last pieces to move in the sequence. So if a bishop was the first piece to move in the sequence, the value for attribute 8 would be '3' and if a knight was played in the last move of the sequence, the value of attribute 9 would be '2'. The tenth and final attribute used refers to an approximation of how spread out the pieces are on the board in the initial position. It is described using the following equation. *s*() denotes the number of pieces in the field of a particular piece, p_n, i.e. in the squares immediately around it. A detailed explanation of the sparsity concept is available in (Iqbal and Yaacob 2008).

$$sparsity = \left[\left(n^{-1} \cdot \sum_{1}^{n} s(p_n)\right) + 1\right]^{-1}$$

A screen capture showing an excerpt of the spreadsheet with DSNS strings on the left and right (150 on each side) and the deviation value in the middle is available in Appendix B. A 'random' composing approach was also used as a control. Essentially, this approach does not use any 'technology' in composing chess problems and places the pieces on the board purely at random. Details are available in (Iqbal 2011). CHESTHETICA was allowed to automatically compose using this DSNS approach (and the random one) for a total of 24 h, i.e. 12 h on one machine and another 12 h on another. The first computer (PC 1) was a notebook: Intel(R) Core(TM) i7-3820QM CPU @ 2.70 GHz with 16 GB of RAM running Microsoft Windows 7 Pro SP1 64-bit. The second computer (PC 2) was a desktop: Intel(R) Core(TM) Duo CPU E8400 @ 3.00 GHz with 4 GB of RAM running Microsoft Windows 7 Pro SP1 32-bit. Table 4.1 shows the 'real world' performance of CHESTHETICA in generating compositions using the DSNS approach.

The average composing rate (using the DSNS approach) for PC 1 was 5.27 compositions per hour (cph) whereas for PC 2 it was 4.23 cph. The sample sizes are too small, at this point, for statistical analysis but it is fair to assume that processing power and memory do not make very much of a difference in this case. Table 4.2 shows the mean aesthetics scores and variations for the (total) generated compositions derived from each set. The higher the number of variations, the 'richer' or 'more complex' the composition is typically regarded to be, according to convention. Despite the precision of these aesthetics scores they are typically used for *ranking* purposes. One should therefore not make claims that an aesthetics score of 2.0 implies that a composition is *twice* as beautiful as one with an aesthetics score of 1.0. At the same time, a small yet statistically significant difference in mean aesthetics (between groups) is not necessarily meaningless. A single factor analysis of variance (ANOVA) test was performed across all the five sets comparing the aesthetics means and the differences were found to be statistically significant: $F(4, 667) = 4.33$, $p = 0.0018$. Even excluding the 'random approach' control, the differences in means were still statistically significant: $F(3, 452) = 3.14$, $p = 0.0251$.

So we see a clear distinction in mean aesthetic value between compositions generated from higher quality sources (Comp3.5, TG2500) and lower quality sources (Comp1.25, TG1500) using the DSNS approach. Interestingly, compositions generated using higher quality tournament games fared slightly better, aesthetically, than low quality compositions. The random approach, expectedly, fared the poorest aesthetically. There was no statistically significant difference in the

Table 4.1 Compositions generated by CHESTHETICA using the DSNS approach (one domain) in 24 h total

	Sources for DSNS approach				Random approach
	Comp3.5	Comp1.25	TG2500	TG1500	
PC 1	30	106	39	78	110
PC 2	41	108	17	37	106
Total	71	214	56	115	226

Table 4.2 Mean aesthetics scores and variations for the compositions generated (one domain)

	Sources for DSNS approach				Random approach
	Comp3.5	Comp1.25	TG2500	TG1500	
Aesthetics score	2.438	2.278	2.308	2.278	2.208
Variations	49.9	31.9	30.0	22.0	35.5

mean variations across all the five sets. This supports previous work that suggested variations are relevant but only to a point (Iqbal et al. 2012). In other words, simply having *more* variations does not necessarily imply higher aesthetic quality.

4.4 The Second Experiment: Different Domains

The first experiment (see previous section) demonstrated that the DSNS approach can indeed produce compositions of higher creative value (than a purely random approach) based on the sources used for the DSNS strings. In the second experiment, we tested to see if materials sourced from *different* domains could also be used successfully. The four domains included: renowned human artworks (i.e. paintings), computer-generated abstract art pieces (from the Elvira system), photographs (with people in them, but not selfies) and renowned classical music pieces. The paintings, photographs and music were identified and selected (300 objects each) by a female research assistant[4] who is also a co-author. The abstract art pieces (1000 in total) were supplied by another two co-authors and from these, the first 300 based on the file names were used. These pieces were created using the Elvira system (Colton et al. 2011). The selection process of these objects were considered sufficiently random for our purposes.

The paintings and classical music pieces were considered high quality sources in contrast to the abstract art pieces that were computer-evolved and considered of comparatively lower quality. The photographs were considered to be of 'moderate' quality; perhaps somewhere between the computer-evolved abstract art pieces and the renowned paintings. Each of these four sets of 300 objects were divided into subsets of 150 as in the first experiment. The same DSNS approach was then used; refer to Chap. 3, Sect. 3.3 for information about merging deviations from different domains. The attributes used for the paintings, photographs and abstract art pieces are as follows. As with the chess domain they represent unique features that a human observer might notice or be able to find out. Similarly they are arbitrary but should be obtainable by some means and describable using real numbers.

1. Resolution (i.e. the total number of pixels)
2. Number of colors used in the image
3. Number of distinct objects in the image
4. The year it was created
5. Aspect ratio (i.e. width divided by height)
6. Brightness
7. Contrast
8. Noisiness
9. Lightness
10. File size (in bits)

[4]We specify the gender of the selector here in case it is ever found to have influenced the results.

The third attribute was determined through 'manual' and direct observation of each image by our female research assistant. These included, for example, people, tables and vases; so two people, one table and three vases would equal six distinct objects. 'Brightness' is a reference to how much all the pixels go from black (dark) to white (bright) (Bezryadin et al. 2007). The 'contrast' value was calculated as explained in (Koren 2006). 'Noisiness' relates to how random or unrelated, pixels are with other pixels surrounding them (Kerr 2008). 'Lightness' as in HSL (or Hue, Saturation, Lightness) was calculated as explained in (Yu et al. 2006). Much of this attribute information was obtained through programmatic means by our other (male) research assistant who is also a co-author. As with the previous experiment, attributes that can be used to describe the object are arbitrary as long as they can be represented using real numbers. Figures 4.2 and 4.3 show examples of the renowned paintings and computer-evolved abstract art pieces used, respectively.

The attributes used for the classical music pieces comprised of the following.

1. Number of channels
2. Sample rate/frequency
3. File size (in bytes)
4. The year it was composed
5. Average loudness
6. Sound energy
7. Dynamic range
8. Bitrate
9. Sample size/quantization
10. Sound efficiency

'Average loudness' refers to the average peak to peak amplitude as calculated using the Python programming language *Average (audio file)* function. 'Sound energy' refers to the zero-crossing rate that the sound goes through, i.e. the amplitude going to zero. The greater the zero-crossing rate, the more energy it

(a) (b)

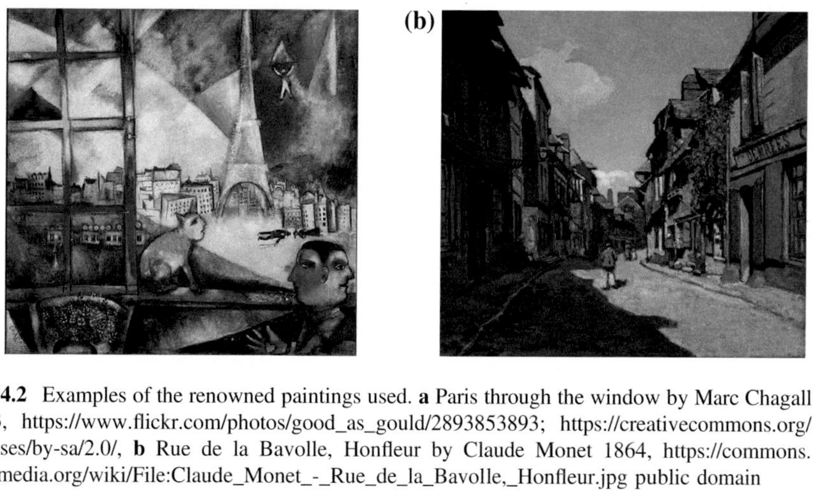

Fig. 4.2 Examples of the renowned paintings used. **a** Paris through the window by Marc Chagall 1913, https://www.flickr.com/photos/good_as_gould/2893853893; https://creativecommons.org/licenses/by-sa/2.0/, **b** Rue de la Bavolle, Honfleur by Claude Monet 1864, https://commons.wikimedia.org/wiki/File:Claude_Monet_-_Rue_de_la_Bavolle,_Honfleur.jpg public domain

(a) (b)

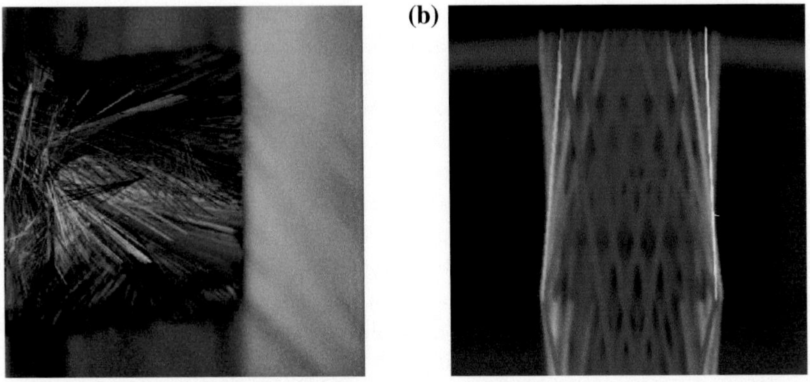

Fig. 4.3 Examples of the computer-evolved abstract art pieces used. **a** *Byline* by the Elvira system, 2010, supplied by co-author Simon Colton and his research associate Cameron Browne, **b** *Crown* by the Elvira system, 2010, supplied by co-author Simon Colton and his research associate Cameron Browne

produces (Srinivasan et al. 1999). The 'dynamic range' is basically the noise floor subtracted from the peak signal. The 'bitrate' is the number of bits of data used per second (e.g. 128 kbps) in the audio file. The 'sample size' refers to the quantization; typically 8 or 16 bits; 24 bits is also possible but less common. The 'sound efficiency' refers to the effectiveness of the sound signal measured by its crest factor, i.e. a measure of the impulsiveness of a noise or vibration signal (Norton and Karczub 2003). Examples of the classical music pieces used include: *Handel—Sarabande*, *Shostakovich—Romance* and *Mozart—Lacrimosa*.

Table 4.3 Compositions generated by CHESTHETICA using the DSNS approach (two domains) in 24 h total

	Comp3.5	Comp1.25	TG2500	TG1500
+Art	37	75	75	65
	38	94	94	50
	75	**169**	**169**	**115**
+Elvira	46	128	128	51
	38	90	90	44
	84	**218**	**218**	**95**
+Photo	38	94	94	36
	32	88	88	33
	70	**182**	**182**	**69**
+Music	35	81	81	58
	42	98	98	57
	77	**179**	**179**	**115**

Once again, for brevity, the chess source sets will be referred to as Comp3.5, Comp1.25, TG2500 and TG1500. The other domain sets with which the chess sets are merged with will be referred to as *art* (renowned paintings), *Elvira* (evolved abstract art pieces), *photo* (photographs of people) and *music* (renowned classical music pieces). In Table 4.3, for each pairing, the topmost row represents the number of compositions generated by CHESTHETICA for PC1, the middle row PC2 and the bottom row, in bold, is the total. So, for example, the Comp3.5 domain merged with the Elvira domain yielded 84 generated compositions in total.

In principle, PC 1 is more powerful than PC 2 in terms of the processing power and available memory. In the previous experiment, there were too few samples to present any statistical analysis of the results. However, combining the information related to the DSNS approach in Table 4.1 with the information in Table 4.3, we have a sample size of 20 each for PC 1 and PC 2, which is not unreasonable for statistical purposes. A two-sample F-test for variances was first applied to the samples to determine whether a two-sample, two-tailed, T-test assuming equal or unequal variances (i.e. TTEV or TTUV) should be used to compare the means (at the 5 % significance level), i.e. the average number of compositions generated in 12 h on a particular machine. This approach will henceforth be used where the means of two samples need to be compared.

It turns out that, based on a TTEV test, there was no statistically significant difference between the average number of compositions composed on PC 1 (worked out to a rate of 5.729 cph) and PC 2 (a rate of 5.321 cph). This finding is important because some readers may assume all of this is simply brute force calculation related directly to how fast a computer can process data and how much memory it has to work with. No doubt, these *are* factors but it does not necessarily indicate the output potential of the DSNS composing approach. A useful analogy may be assuming that simply turning up the heat in a room will dry wet clothes proportionately faster. Table 4.4 shows, for each pairing, the mean aesthetics scores (top row) and mean variations (bottom row).

For Comp3.5, an ANOVA test across 5 sets, i.e. Comp3.5 on its own (2.438 mean aesthetics score, see Table 4.2) and Comp3.5 merged with the four other domains (see Table 4.4), showed a statistically significant difference between the means for their aesthetics scores: $F(4, 372) = 2.69$, $p = 0.0031$. This suggests that, given a high quality source (Comp3.5), using 'information' from other domains to compose chess problems does not result in higher quality compositions since the

Table 4.4 Mean aesthetics scores and variations for the compositions generated (two domains)

	Comp3.5	Comp1.25	TG2500	TG1500
+Art	2.243	2.289	2.250	2.363
	63	33.6	28	19.8
+Elvira	2.277	2.328	2.352	2.330
	41.2	43.2	58.3	24.6
+Photo	2.323	2.304	2.137	2.449
	35.9	33.6	21.5	44.2
+Music	2.270	2.302	2.364	2.400
	63	30.2	60.8	31.2

aesthetics scores using two domains were all lower. There was no difference of statistical significance for the average number of variations between the sets, which is the same outcome as the previous experiment.

For Comp1.25, a similar ANOVA test showed no statistically significant difference between the means for their aesthetics scores or the average number of variations between the sets. This suggests that given a *lower* quality chess source (2.278 mean aesthetics score for Comp1.25 alone, see Table 4.2), using information from other domains neither negatively nor positively affects the quality of the generated compositions. For TG2500, however, the same ANOVA test showed a statistically significant difference between the means for their aesthetics scores: $F(4, 327) = 3.48$, $p = 0.0084$, but not for the variations. This suggests that using a high quality source of *tournament game sequences* (2.308 mean aesthetics score for TG2500 alone, see Table 4.2), the quality of the generated compositions can actually be improved slightly if used in combination with the Elvira or classical music sets. Yet the opposite is true if combined with the art and photo domains in this case.

Finally, for TG1500, the ANOVA test showed a statistically significant difference between the means for their aesthetics scores: $F(4, 504) = 2.39$, $p = 0.0497$; and *also* for the variations: $F(4, 504) = 2.51$, $p = 0.0412$. This suggests that using a *low* quality set of tournament game sequences (2.278 mean aesthetics score for TG1500 alone, see Table 4.2), the quality of the generated compositions can actually be improved if used in combination with *any* of the four other domains. In contrast with the result obtained for Comp1.25 above, the result also suggests that there is something 'intrinsically' different about compositions and tournament game sequences given that, in this case, their mean aesthetics scores using a single domain source were exactly the same (i.e. 2.278). If, in principle, we can accept that compositions, even low quality ones, are generally better than low quality tournament game sequences, then the results so far would suggest that the *lowest* quality source domain (i.e. TG1500) used in combination with other domains actually produces among the best results.

Even though the difference in the average number of variations in the case of TG1500 was also statistically significant, it is difficult to prove that having more or fewer variations (and what the thresholds really are) improves the quality of a composition. We can only say that the DSNS approach, in combining information from other domains with low quality tournament game sequences, is able to influence the average number of variations for the compositions generated both positively and negatively. Returning to the mean aesthetics scores, it is not clear why, exactly, the lowest quality chess source used in combination with *all* the other completely unrelated domains tested, through the DSNS approach, should produce compositions of higher quality than using the low quality chess source alone. It suggests that 'unrelated' information from different domains can indeed be aggregated meaningfully as is likely occurring in some form in the human brain.

To test this further, we performed another comparison of three sets involving the TG1500 and photo domains. This is because the 'TG1500 + photo' combination (TG1500p for brevity) yielded the highest, statistically significant mean aesthetics score (2.449). We introduced a new source material set, i.e. 'photorandom'. This is basically the same set of 300 photographs of people except that the values for each

Table 4.5 Mean aesthetics scores and variations for the compositions generated (one and two domains)

	TG1500	TG1500p	TG1500pr
Aesthetics score	2.278	2.449	2.325
Variations	22.0	44.2	19.6

attribute were randomly-generated within the range of highest and lowest for that attribute. So, for instance, if the 'number of colors' attribute for all 300 photos ranged from 927 to 238,034, then a 'random' photo would have a new value for that attribute based on a random number generated within that range. This was done for all 10 attributes. The result was a collection of 'garbage' photos where the attributes of each made no sense and probably did not exist in reality.

It stands to reason that, if information is indeed being 'successfully' integrated or aggregated between unrelated domain types using the DSNS approach, *actual* photos (with *real* attribute values) should do better than those with randomly-generated attribute values (garbage photos) and also better than *not* using photos at all. So we compared the TG1500 (alone with no photo integration), the TG1500p *and* the TG1500 + photorandom (TG1500pr for brevity) sets together. The results are shown in Table 4.5.

An ANOVA test showed a statistically significant difference between the means for their aesthetics scores: $F(2, 286) = 3.62$, $p = 0.0281$; and also for the variations: $F(2, 286) = 3.60$, $p = 0.0286$. So it would seem that using actual photos (with real attributes) performs better than garbage photo data and also better than not using photos at all. Using actual photos also yields the highest number of variations, on average, of the three sets but that alone is not necessarily an improvement in terms of quality because we can see that the non-DSNS random approach in Table 4.2 also had a relatively high number of variations (35.5) but the lowest mean aesthetics score (2.208) compared to the other single domain chess sources. In summary, the DSNS approach, at least in the case of low quality tournament game sequences integrating information from real photographs, is able to produce compositions of higher aesthetic quality than otherwise. It remains an open question for now why this should happen at all, even though the human brain likely successfully mingles information from different domains in unusual and poorly-understood ways as well.

4.5 The Third Experiment: Variations in the Number of Objects and Attributes

In the third experiment, we wanted to test if the number of *objects* used in the DSNS approach and the number of *attributes* influenced the quality of the compositions generated. For this purpose, we used the Comp3.5 and TG1500 samples contrasted against variations of each where only 150 objects were used (instead of 300) and also against the same sample size but where only 5 attributes were used,

Table 4.6 Mean aesthetics scores and variations for the compositions generated (one high quality source, variable objects and attributes)

	Comp3.5 300 objects 10 attributes	Comp3.5 150 objects 10 attributes	Comp3.5 300 objects 5 attributes
Aesthetics score	2.438	2.263	2.272
Variations	49.9	50.6	26.3

Table 4.7 Mean aesthetics scores and variations for the compositions generated (one low quality source, variable objects and attributes)

	TG1500 300 objects 10 attributes	TG1500 150 objects 10 attributes	TG1500 300 objects 5 attributes
Aesthetics score	2.278	2.257	2.335
Variations	22.0	22.1	29.3

i.e. the first 5 attributes (see Sect. 4.3). The automatic composing process (see Appendix A) would simply skip the constraints that depended upon the missing attributes. Comp3.5 and TG1500 were used because these had, respectively, the highest and lowest aesthetics scores (see Table 4.2). The results are shown in Tables 4.6 and 4.7.

For Comp3.5 (300 objects, 10 attributes) against Comp3.5 (150 objects, 10 attributes), the difference between the mean aesthetics scores based on a TTEV test was statistically significant: $t(148) = 2.439$, $P = 0.0159$. For Comp3.5 (300 objects, 10 attributes) against Comp3.5 (300 objects, 5 attributes), the difference between the mean aesthetics scores based on the same test was statistically significant as well: $t(169) = 2.749$, $P = 0.0033$. This suggests that, given a high quality chess source, using more objects or using more attributes improves the quality of the generated compositions, which implies scalability of the DSNS approach. There were no differences of statistical significance (TTUV) between the average number of variations comparing the same groups.

For TG1500 (300 objects, 10 attributes) against TG1500 (150 objects, 10 attributes), the difference between the mean aesthetics scores based on a TTEV test was not statistically significant. For TG1500 (300 objects, 10 attributes) against TG1500 (300 objects, 5 attributes), the difference between the mean aesthetics scores based on a TTUV test was not statistically significant either. This suggests that using more objects or more attributes makes no difference given a low quality chess source. There were no differences of statistical significance (TTUV) between the average number of variations comparing the same groups. We also tested a merged set, i.e. TG1500p (see previous section) using just 150 objects and 5 attributes as follows (Tables 4.8 and 4.9). For the photos, we used the first 5 attributes (see Sect. 4.4).

Table 4.8 Mean aesthetics scores and variations for the compositions generated (two domains, variable objects)

	TG1500p 300 games 300 photos 10 attributes	TG1500p 150 games 150 photos 10 attributes	TG1500p 150 games 300 photos 10 attributes	TG1500p 300 games 150 photos 10 attributes
Aesthetics score	2.449	2.171	2.363	2.306
Variations	44.2	24.5	39.9	35.1

Table 4.9 Mean aesthetics scores and variations for the compositions generated (two domains, variable attributes)

	TG1500p 300 games 300 photos 10 attributes	TG1500p 300 games 300 photos 5 attributes
Aesthetics score	2.449	2.355
Variations	44.2	25

For TG1500p (300 games, 300 photos, 10 attributes) against TG1500p (150 games, 150 photos, 10 attributes), the difference between the mean aesthetics scores based on a TTEV test was statistically significant: t(153) = 3.779, P = 0.0001. For TG1500p (300 games, 300 photos, 10 attributes) against TG1500p (150 games, 300 photos, 10 attributes), the difference between the mean aesthetics scores based on the same test was not statistically significant. For TG1500p (300 games, 300 photos, 10 attributes) against TG1500p (300 games, 150 photos, 10 attributes), the difference between the mean aesthetics scores based on the same test was statistically significant: t(204) = 2.219, P = 0.0276. So using more *objects* for both sources given these two source domains improves the quality of the compositions generated. This is consistent with the previous finding when using only a high quality chess source (Comp3.5, Table 4.6). However, for TG1500p sources, reducing just the number of games does not make a difference; but reducing the number of *photos*, interestingly, does. There were no differences of statistical significance across these four sets between the average number of variations based on an ANOVA test.

For TG1500p (300 games, 300 photos, 10 attributes) against TG1500p (300 games, 300 photos, 5 attributes), the difference between the mean aesthetics scores based on a TTUV test was not statistically significant. This suggests that, given these two source domains, using more *attributes* makes no difference. This is somewhat consistent with the previous finding when using only a low quality chess source (TG1500, Table 4.7), in that an increase in the number of *attributes* used does not make a difference, regardless of the inclusion of photos as a source. There were no differences of statistical significance between the average number of variations based on a TTUV test.

4.6 The Fourth Experiment: Human Expert Assessment

As explained in Sect. 4.2, enlisting the help of human experts to evaluate aesthetics in this research would have its share of problems. Fatigue, imprecision and inconsistency are hallmarks of the human condition especially when it comes to typically subjective issues such as aesthetics. Nonetheless, if the amount of work required can be kept manageable and interesting for the experts, it is still useful to see what their evaluations might be like. So in the fourth experiment, we used the compositions generated using three approaches, i.e. TG1500p, TG2500 + photo (TG2500p for brevity) and Comp3.5. The following positions (see Table 4.10) were excluded from the compositions generated in all the sets because they are too simplistic and common to retain the attention of the human experts. K = King, Q = Queen, R = Rook, P = Pawn. A digit before the alphabet indicates the piece count.

From the remainder, the top 30 from each set based on their aesthetics scores were chosen, and these randomly mixed together into a PGN (Portable Game Notation) file or database of 90 compositions. These DSNS-generated compositions and their solutions are provided in Appendix C. All identifying information was also removed from each composition in the file so the human experts would not know which set they came from, or even if they were composed by a human or computer. For a balanced view of aesthetics in the game, the human experts chosen were co-authors Matej Guid (FIDE[5] Master), Jana Krivec (Woman Grandmaster) and Vlaicu Crisan (International Master of Chess Solving and FIDE Master of Composition). Together they represent more or less the spectrum of expertise pertaining to the game of chess and where aesthetics in the game has been recognized, i.e. over the board and also in the domain of composing chess problems. They were considered sufficiently 'domain competent' with regard to the game. All the experts were completely unaware of the details of this fourth experiment until it was completed.

The human experts were asked to rate the 90 compositions on a scale of 0.0–5.0 (one precision point) "*based on their individual expertise and perception of beauty/aesthetics in the game*". This would also mean that some compositions would have to have the same rating based on human perception, which is acceptable. A larger scale of say, 0–10 was not used because this has already been attempted in previous research work related to the aesthetics model (Iqbal 2008; Iqbal et al. 2012) and we wanted to try something different. Comments for each composition were optional. After rating the 90 compositions, the experts were asked to then choose 15 from the 90 that were "*most likely to have been composed by a human being*". The relevance of this second part will become evident later. The PGN file and Microsoft Excel evaluation sheet were e-mailed to the experts independently and they were asked to respond within two weeks. One expert took slightly longer due to other commitments. This is understandable given the complexity of having to (comparatively)

[5]Fédération Internationale des Échecs (World Chess Federation).

Table 4.10 Composition types that were excluded for the benefit of the human experts

K, Q versus K	K, R versus K	K, 2R versus K
K, Q versus K, P	K, R versus K, P	K, 2R versus K, P
K, Q, P versus K	K, R, P versus K	K, 2R, P versus K
K, Q, P versus K, P	K, R, P versus K, P	K, 2R, P versus K, P
K, Q, R versus K		
K, Q, R versus K, P		

rate 90 compositions on a discrete scale based on the subjective aspect of beauty. Their ratings are shown in Appendix D. Two of the human experts provided optional commentary about the merits and issues pertaining to each composition and this is provided in Appendix E.

In general, the experts were of the opinion that the compositions they were given were of low quality or too easy. This is to be expected given the filter of just one composition convention (see Sect. 4.2) and possibly because the automatic composer 'reduces' each problem to be as economical as possible (see Appendix A). Imposing all five composition conventions using CHESTHETICA would have taken far too long for our experimental purposes. One expert—the master solver and composer—suspected all the compositions were generated by computer but then went on to choose 16 instead of 15 that he thought were composed by a human. The female grandmaster selected 27 instead of just 15 whereas the FIDE master kept to the instructed 15. This alone suggests that the compositions generated, by expert opinion, were more human-like than to be expected; especially given that they were indeed *all* generated by computer using the DSNS approach. In fact, the experts did tend to have nicer things to say about the ones they thought were composed by a human (see Appendix E).

The average expert score for the 90 compositions they evaluated were assessed based on the original three source sets (which the experts had no idea about), i.e. Comp3.5, TG1500p and TG2500p. An ANOVA test showed no statistically significant difference between the means (i.e. 0.847, 0.722 and 0.776, respectively). This was on a scale of 0.0–5.0. The result could have been due to the small sample sizes and the overall difficulty of the task in discerning between many compositions that may appear quite similar in quality, if not form as well. There was also no statistically significant (Pearson) correlation between any two of the human expert evaluations. This was not entirely unexpected as humans tend to factor in personal tastes and biases often without realizing it (Iqbal 2014).

The same sets were tested using CHESTHETICA and an ANOVA test showed a statistically significant difference between the means (i.e. 2.678, 2.538 and 2.343, respectively): $F(2, 87) = 6.154$, $p = 0.0032$. So the human experts were unable to discern between the groups based on their average aesthetics ratings but the computer program seemed to think that Comp3.5, given this subset of 30 compositions, produced the highest quality chess problems compared to TG1500p and TG2500p, with the former doing better than the latter. This contradicts the means for these groups (i.e. 2.438, 2.449 and 2.137, respectively) shown in Tables 4.2 and

4.4 which had larger sample sizes because an ANOVA test showed the differences in means here to be significant as well: $F(2, 214) = 12.358$, $p = 8.3E\text{-}06$. Here, TG1500p is slightly better than Comp3.5, and TG2500p is the worst of the lot.

The deciding factor, we thought, must reside in subtle form somewhere in the 'undecided' human expert aesthetic evaluations. Looking back at them, the suggestive answer was actually in their selections of the compositions *thought to have been composed by a human* (which also tended to receive the most favorable comments). Considering only the compositions where *two* or more experts agreed—classified in terms of the sources used to produce them—the experts' selections tended to agree with CHESTHETICA's assessment just mentioned. See Table 4.11.

The final standings of the compositions thought to have been composed by a human when tallied in terms of the number of experts (in cases where the majority agreed) stood at: TG1500p (8), Comp3.5 (7), TG2500p (2). This correlates perfectly with the order of CHESTHETICA's assessment of the mean aesthetics scores earlier using all available compositions for these sets, i.e. 2.449, 2.438 and 2.137, respectively. There was absolutely no way the experts could have known the compositions they were selecting favorably had come from these particular source sets. The rough probability of any two experts agreeing with each other that any particular composition was composed by a human = $[[(15 + 16 + 27)/3]/90]^2 \times 100 = 4.61\ \%$. The rough probability of all *three* experts agreeing with each other on a particular composition is 0.99 %. Figure 4.4 shows two of the DSNS-generated compositions the majority of experts agreed upon were composed by a human being. The solutions are in Appendix C.

So it would seem that, based on the available evidence, TG1500p does indeed, on average, produce compositions that are *slightly* better in quality than Comp3.5 alone; and that TG2500p is indeed the worst of the three in terms of average composition quality. This suggests that real (i.e. actual) photographic information, used in conjunction with a low quality chess source (TG1500) via the DSNS approach is able to produce creative objects in the domain of chess of comparable quality or even slightly better than using a high quality chess source alone (Comp3.5). Again, the reason for this remains an open question for now.

Table 4.11 Expert agreement on the computer compositions thought to be human compositions

#	Experts agreed	Source
24	2/3	Comp3.5
44	2/3	TG1500 + photo
55	2/3	TG2500 + photo
62	2/3	TG1500 + photo
63	2/3	TG1500 + photo
66	2/3	Comp3.5
81	3/3	Comp3.5
88	2/3	TG1500 + photo

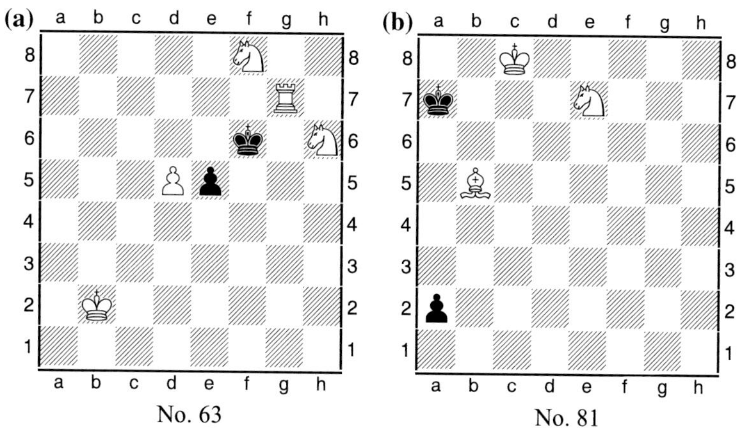

Fig. 4.4 Examples of the DSNS-generated #3 compositions

4.7 The Fifth Experiment: Comparing TG1500p and Comp3.5

To be sure, an experiment was performed to gauge the overall performance differences between the TG1500p and Comp3.5 approaches. This time, three different computers were used, i.e. PC1 and PC2 (see Sect. 4.3) and PC3, a notebook: Intel (R) Core(TM) i3 CPU M 370 @ 2.40 GHz with 4 GB of RAM running Microsoft Windows 7 Home Premium 64-bit. Each computer had different processing power and capabilities; in general, in reducing amount from PC1 to PC3. CHESTHETICA was set to run for 24 cycles of automatic composition on each computer, comparing these two approaches. Each cycle was set for 6 h and the composing efficiency in compositions per hour (cph) was measured per cycle. The composing process was this time tested using two convention filters (i.e. *no checks in key move* and *no captures in key move*) and also using no convention filters; the former was expected to produce compositions of higher quality but at reduced efficiency.

Compiling the composing efficiency readings from all 72 cycles, we found that the average *cph* using two convention filters based on the TG1500p approach (i.e. 3.32 cph) was indeed higher than the Comp3.5 approach (i.e. 2.34 cph) to a statistically significant degree based on a TTUV test: $t(108) = -4.406$, $P < 0.0001$. However, using no convention filters the TG1500p approach (i.e. 8.97 cph) was indeed lower than the Comp3.5 approach (i.e. 10.58 cph) based on a TTEV test: $t(122) = 2.924$, $P = 0.004$. The consolidated aesthetic scores for all of the compositions produced during those 72 cycles showed that there was no statistically significant difference between TG1500p (i.e. 2.360) and Comp3.5 (2.343) using two convention filters but there was when using none, i.e. 2.3 and 2.271, respectively, based on a TTUV test: $t(8415) = -3.262$, $P = 0.0011$. We also confirmed, as predicted, that using two convention filters, the aesthetics scores do, on average, improve compared to using no filters. This was true for both Comp3.5 (two filters: 2.343, none: 2.271) and TG1500p (two filters: 2.36, none: 2.3) based on a TTUV and TTEV test, respectively: $t(1666) = 5.322$, $P < 0.0001$ and $t(5393) = 4.828$, $P < 0.0001$.

So in summary, for more beautiful compositions (i.e. using two convention filters), the TG1500p approach is more efficient than the Comp3.5 approach. It produces more compositions in the same amount of composing time. Comp3.5 produces comparable compositions (in terms of aesthetic quality) but fewer. Using no convention filters, Comp3.5 produces more compositions than TG1500p but of lower aesthetic quality. So the TG1500p is clearly the better of the two approaches. Somehow, combining information from weak games and photographs of people (using the DSNS approach) outperforms using information from high quality chess compositions alone. Interestingly, based on ANOVA tests, there was also no statistically significant difference in average composing efficiency between the three computers used in all cases but one (i.e. TG1500p, two convention filters), where the weakest computer (PC3) actually performed the best: $F(2, 69) = 3.1475$, $p < 0.05$. Perhaps an indicator of a machine employing 'genuine' creativity is a demonstration that raw processing power does not necessarily improve performance; and that perhaps even a relatively low power machine can outperform higher power ones. In principle, creativity largely happens when it happens and can only be improved using a better approach or process rather than a faster machine.

4.8 The Sixth Experiment: Comparisons Against the State-of-the-Art

The state-of-the-art with regard to a computational creativity approach and chess problem composition, to the best of our knowledge, is the 'experience table' described in (Iqbal 2011). Essentially, it uses domain knowledge extracted from a database of 29,453 mostly published chess compositions by human composers, many of whom are of the expert and master level. The experience table is a table of probabilities that a piece of a particular type should exist on a particular square on the board. This information is used in the automatic composing process and was shown to outperform a random-placement approach (similar to the one used in Sect. 4.3) and also experience tables derived from other 'weaker' sources. Full details are in the aforementioned reference.

So in order to compare the DSNS approach with the experience table one, we used three pairings of chess three-movers and photographs for the DSNS, and the original 29,453 source database for the 'experience table' approach. We used three pairings of the DSNS because there are many different ways even the best pairing can be represented, i.e. it is not limited to a particular set or size of chess sequences or photographs. The three pairings therefore provide a more balanced view of the

DSNS. The experience table, on the other hand, was shown to perform best using the aforementioned dabatase so we used only that. The three DSNS pairings included the original TG1500p (as explained in Sect. 4.4), and two others, i.e. 4698 games between weak players with an Elo rating below 1600 in conjunction with 1000 photographs of people (TG1600p1 K for brevity), and TG1500 (see Sect. 4.3) in conjunction with the same 1000 photographs of people (TG1500p1 K for brevity).

We tested using no chess composition conventions, using two conventions (*no checks in key move* and *no captures in key move*) and using three conventions (the previous two and *no cooked problems*). In general, the more conventions applied, the higher the quality of the chess problems (Iqbal 2014). So for the tests we used one on the low end (i.e. no conventions) and two tests on the higher end. Table 4.12 shows the mean aesthetics scores of the compositions generated and the composing efficiency in compositions per hour (cph) based on 24 cycles (as explained in Sect. 4.7). For three conventions, we used 36 cycles because it takes longer to compose a reasonable amount of problems for statistical analysis.

For no conventions, we used the TTEV test (see Sect. 4.4) because the sample sizes were too large but for the rest we used the Mann-Whitney U test (two-tailed, 5 % significance level) to compare the means, which is suitable if the sample sizes are smaller, i.e. typically below 200. There was no statistically significant difference between any of the three paired sets (in terms of both aesthetics and efficiency) except in the case of the mean aesthetics scores when applying *three* conventions (U-value = 7382.5, Z-score = −2.1385, n_1 = 121, n_2 = 144, P = 0.03236). This suggests that, under stricter composing conditions, the DSNS approach outperforms the experience table approach by producing compositions that, on average, rank higher aesthetically. Furthermore, the composing rate is still statistically equivalent. Under less strict composing conditions, there is no difference between the two approaches. Even then, it still leaves something to be said about the DSNS considering that the experience table approach uses tens of thousands of published compositions by human composers as a source whereas the DSNS uses sequences from games between weak players and the entirely unrelated domain of photographs of people.

Table 4.12 Comparison of the DSNS against the state-of-the-art 'experience table' approach

Conventions	None		Two		Three	
Composing approach	DSNS: TG1500p	Experience table	DSNS: TG1600p1K	Experience table	DSNS: TG1500p1K	Experience table
Mean aesthetics score	2.194	2.218	2.285	2.257	2.316	2.240
Mean efficiency (cph)	4.709	4.119	0.799	0.993	0.555	0.667

4.9 Human Versus Computer Composition

For the interested reader, in terms of mean aesthetics (as assessed by the aesthetics model and perceived by the majority of domain-competent human players), there was no statistically significant difference between the 69 compositions created using the TG1500p DSNS approach (2.449) and 69 randomly-selected compositions by human experts from The Meson Chess Problem Database (2.300) explained in Sect. 4.3; TTEV test. There was, however, a statistically significant difference between the average number of variations where TG1500p had 44.2 variations on average and the Meson sample 143.1 on average; TTUV test: $t(93) = 2.937$, $P = 0.00418$. Again, it is important to note that chess problems have various types (e.g. three-movers, endgame studies, constructs) and different requirements which go beyond the common ground of aesthetics assessed by the aesthetics model.

For instance, compositions by humans tend to feature more variations by design and there is insufficient evidence that simply having more variations leads to improved aesthetics. Likewise, a complex yet award-winning problem may be difficult for the *majority* of chess players to perceive as beautiful whereas an unexpected yet direct tactical checkmate (e.g. from a famous tournament game or construct) might be easier to fathom and more appealing to them. So the comparison between the TG1500p and Meson Database samples suggest that, on average, *aesthetically*, at least, the computer-generated compositions are comparable to (but not better than) those created by human experts. A growing collection of computer-generated chess problems composed using the DSNS approach is available online[6] for interested readers. A selection has also been published in a chess problem magazine (Enemark 2015). In addition to being appreciated aesthetically, they can also be used by players of all levels as puzzles to train and improve their game. Additionally, between 5 January and 19 February 2015, for instance, 1263 three-movers were composed in total using anywhere between one and nine different instances of CHESTHETICA and no repetitions were detected. The actual implementation in code of the most recent version of Chesthetica's composing function is given in Appendix F, as the simplified algorithm in Appendix A may be insufficient for programmers intending to develop a similar working system.

[6]Search YouTube for the channel "Azlan Iqbal" or "Chesthetica" or go to https://www.youtube.com/c/AzlanIqbal.

References

Bezryadin S, Bourov P, Ilinih D (2007) Brightness calculation in digital image processing. In: International symposium on technologies for digital photo fulfillment, vol 2007, no 1, pp 10–15. Society for Imaging Science and Technology

Colton S, Cook M, Raad A (2011) Ludic considerations of tablet-based evo-art. Applications of evolutionary computation. Springer, Berlin, pp 223–233

Enemark B (2015) Computer-generated problems, Problemskak, Denmark, vol 4, no 31, pp 1, 3–6. ISSN 1903-0169

Galanter P (2012) Computational aesthetic evaluation: past and future. Computers and creativity. Springer, Berlin, pp 255–293

Hesse H (2011) The joys of chess: heroes, battles and brilliancies. New in chess, 1st edn

Humble PN (1993) Chess as an art form. Br J Aesthetics 33:59–66

Humble PN (1995) The aesthetics of chess: a reply to ravilious. Br J Aesthetics 35(4):390

Iqbal MAM (2008) A discrete computational aesthetics model for a zero-sum perfect information game. Ph.D. thesis, University of Malaya, Kuala Lumpur, Malaysia. https://www.researchgate.net/publication/230855649_A_Discrete_Computational_Aesthetics_Model_for_A_Zero-Sum_Perfect_Information_Game

Iqbal A (2011) Increasing efficiency and quality in the automatic composition of three-move mate problems. In: Anacleto J, Fels S, Graham N, Kapralos B, Saif El-Nasr M, Stanley K (eds) Entertainment computing—ICEC 2011. Lecture notes in computer science, vol 6972, 1st edn, XVI. Springer, Berlin, pp 186–197. ISBN 978-3-642-24499-5

Iqbal A (2012) A computer program to identify beauty in problems and studies, ChessBase News, Hamburg, Germany, 15 Dec. http://en.chessbase.com/post/a-computer-program-to-identify-beauty-in-problems-and-studies

Iqbal A (2014) How relevant are chess composition conventions? In: Barbosa SDJ, Chen P, Cuzzocrea A, Du X, Filipe J, Kara O, Kotenko I, Sivalingam KM, Ślęzak D, Washio T, Yang X (eds) Communications in computer and information science, vol 408. Springer, Berlin, pp 122–131. Online ISSN 1865-0937, Print ISSN 1865-0929

Iqbal A (2015) Computer science and artificial intelligence: "computational aesthetics". In: Encyclopedia Britannica, 11 Feb. Encyclopedia Britannica, Inc., Chicago, USA. http://global.britannica.com/EBchecked/topic/2011991/computational-aesthetics

Iqbal A, Yaacob M (2008) Computational assessment of sparsity in board games. In: Proceedings of the 12th international conference on computer games: AI, animation, mobile, educational and serious games (CGames 2008), Louisville, Kentucky, USA, vol 30, pp 29–33

Iqbal A, van der Heijden H, Guid M, Makhmali A (2012) Evaluating the aesthetics of endgame studies: a computational model of human aesthetic perception. IEEE Trans Comput Intell AI Games: Spec Issue Comput Aesthetics Games. 4(3):178–191. ISSN 1943-068X. e-ISSN 1943-0698

Kerr DA (2008) The ISO definition of the dynamic range of a digital still camera. http://dougkerr.net/Pumpkin/articles/ISO_Dynamic_range.pdf

Koren N (2006) The imatest program: comparing cameras with different amounts of sharpening. In: Electronic imaging 2006, p 60690L. International Society for Optics and Photonics

Lord C (1984–85) Is chess art? Philosophic exchange, vols 15 and 16, pp 117–122

Norton MP, Karczub DG (2003) Fundamentals of noise and vibration analysis for engineers. Cambridge University Press, Cambridge

Nunn J (2015) Ostroda: 39th world solving championship—part 2, ChessBase News, Hamburg, Germany. 20 Aug. http://en.chessbase.com/post/ostroda-39th-world-solving-championship-2

Rachels J (1984–85) Chess as art: reflections on richard reti, philosophic exchange, vols 15 and 16, pp 105–115

Ravilious CP (1994) The aesthetics of chess and the chess problem. Br J Aesthetics 34:285–290

Ritchie G (2007) Some empirical criteria for attributing creativity to a computer program. Mind Mach 17(1):67–99

Shannon CE (1950) Programming a computer for playing chess. Phil Mag 41(314):256–275

Srinivasan S, Petkovic D, Ponceleon D (1999) Towards robust features for classifying audio in the CueVideo system. In: Proceedings of the seventh ACM international conference on multimedia (part 1), pp 393–400. ACM

Turing A (1953) Digital computers applied to games. In: Bowden BV (ed) Faster than thought (Chapter 25) (subsection). Pitman, London

Velimirovic M, Valtonen K (2013) The definitive book: encyclopedia of chess problems—themes and terms (reprinted edition)

Yu Y, Xu D, Chen C, Yu Y, Zhao L (2006) A surface errors locator system for ancient culture preservation. Digital libraries: achievements, challenges and opportunities. Springer, Berlin, pp 360–369

Chapter 5
Consolidation of Results

Abstract With reference to the experimental work in Chap. 4, we summarize the main findings in point form for the benefit of a wider readership. Notably, the fact that processing power and memory do not seem to have a significant effect on composing efficiency using the Digital Synaptic Neural Substrate (DSNS) approach, and how the quality of the compositions is higher compared to a random piece placement approach. In addition, the DSNS happens to be better than the present state-of-the-art 'experience table' approach that is also better at composing than simple random piece placement. Furthermore, in certain cases, variations in the number of photographs or chess games used to seed the DSNS process may be significant. Similarly, so do variations in the number of attributes used that represent the fragments of information taken from the aforemention domains. Possible limitations of the DSNS are discussed, including a brief exploration of its potential applications in other domains and fields.

Keywords Results · Findings · DSNS · Efficiency · Limitations · Applications · Domains · Fields

5.1 General Findings

Bringing all of the experimental data of this research together, we present the following findings regarding the DSNS approach, including some extended findings.

5.1.1 On the DSNS

1. Computer processing power and memory do not have a significant effect on the composing efficiency.

A. Iqbal et al., *The Digital Synaptic Neural Substrate*,
SpringerBriefs in Cognitive Computation, DOI 10.1007/978-3-319-28079-0_5

2. Given the use of just a single domain, the DSNS approach produces higher quality output compared to a completely random approach regardless of the quality of that DSNS source.
3. Given the use of just a single domain, a higher quality source results in higher quality output.
4. Given the use of just a single domain of high quality, combining with another unrelated domain has a negative effect on the quality of the output.
5. Given the use of just a single domain of 'moderate' quality (e.g. TG2500)—considering all chess samples used—combining with another unrelated domain improves the quality of the output in some cases but has a negative effect in other cases.
6. Given the use of just a single domain of low quality (preferably *lowest* quality, e.g. TG1500), combining with another unrelated domain improves the quality of the output in *all* cases.
7. Given the use of just a single domain of low quality (i.e. TG1500), combining with *actual* photographs of people produces higher quality output than garbage photo data and not using photos at all.
8. Given the use of just a single domain of high quality, using more objects or more attributes has a positive effect on the quality of the output.
9. Given the use of just a single domain of low quality, using more objects or more attributes has no effect on the quality of the output.
10. For TG1500p (low quality chess domain used in combination with photo domain), increasing the number of objects for both improves the quality of the output. Reducing the number of photos alone has a negative effect. Reducing the number of attributes of both has no effect.
11. Overall, taking into account composing efficiency and quality using different machines, TG1500p outperforms Comp3.5.
12. The DSNS also outperforms the present state-of-the-art composing approach (i.e. the 'experience table') in terms of average aesthetic quality of compositions generated under stricter composing conditions such as using three convention filters. Under lesser conditions, it is equal.
13. The computer-generated compositions using the TG1500p approach are of comparable aesthetic quality to compositions created by human experts, even though award-winning compositions are also judged by other factors not specifically accounted for by the aesthetics model used.

5.1.2 Extended Findings

1. The number of variations in a composition likely has little to no effect on its quality beyond a limited point.
2. Human expert aesthetic assessment of computer-generated compositions (of not particularly high quality) tends to be inconclusive using a discrete scale. Other

methods such as them selecting those which are thought to have been composed by a human tend to reveal more.

These findings can be explored in more detail by the reader by referring back to Chap. 4. Even so, there appears to be a pattern with regard to the DSNS approach, at least when aimed at creating objects of creative value within the domain of chess. It would seem that if the intended domain (i.e. the one in which objects of creative value are to be created) has a high quality source upon which to draw upon (e.g. Comp3.5), then using objects from unrelated domains may be unnecessary. However, a larger number of such objects and a larger number of attributes that describe each of those objects would likely improve the results.

On the other hand, where a high quality source upon which to draw upon is not available—or only a low quality source is available (e.g. TG1500)—then using objects from other domains (e.g. actual photos of people) somehow improves results. Perhaps even to the point that is comparable to using a high quality source alone or even slightly better. The reason for this is not clear. The only explanation that appears to make some sense at this point is that the DSNS approach must be working in a way that is somewhat analogous to the neuro-chemical substances and reactions that represent objects and experiences of all kinds in our brains. When these representations of objects combine and interact in poorly-understood ways, we sometimes are able to produce new objects that are deemed, 'creative'.

5.2 Limitations of the DSNS and Its Possible Application in Other Domains

While we have shown experimentally that the DSNS not only works but also outperforms other comparable approaches in automatic chess problem composition (of the three-mover variety, at least), there are certain limitations to consider. For instance, the 'process' of attribute or feature selection in a given domain is generally arbitrary, though they must be numerically representable. This usually goes against our scientific inclinations to be very precise and well-defined in our methods. However, we must realize that there is no known 'formula' for creativity and that different brains tend to perceive even the same things somewhat differently. This may be why some people are more creative than others. So the lack of a well-intended yet constraining process of attribute selection as may be found in other areas of artificial intelligence creates some confusion about 'what works' and what does not.

There is also a question of how many objects in a particular domain should be used. The permutations can be endless and the only way to be sure, at least to a certain extent, is to run experiments as we have done in Chap. 4, Sects. 4.3, 4.4 and 4.5. We now know, for instance, that games between weak players and photographs of people can be used to generate relatively good three-move chess problems. However, the experiments do not explain *why* this is so. Is there a reason? Perhaps.

Does it matter to scientists and engineers? Probably not because they can still study and build DSNS systems that work and create both artful and useful things (like chess problems). If we found out *why*, could we improve or optimize the DSNS process? Possibly. So the lack of knowing why can be seen as an additional limitation but perhaps just a temporary one. A lot more work by a lot more people in many different areas (e.g. neuroscience, AI, psychology, philosophy) is clearly required to make significant headway in answering the 'why' question.

Can the DSNS be applied in other areas? In principle, the approach itself does not constrain it to chess. If we have or can imagine an existing system that is able to construct say, paintings, given a set of inputs like the number of colors, the size of the canvas, the frequency of a particular kind of stroke and the intensity of certain colors, the DSNS can indeed supply various combinations of these numbers that might result in interesting paintings. There is no way to be sure unless detailed and rigorous experiments, like the ones in this book, are done. Perhaps drawings by young children in combination with classical music would produce the best results. So any domain that is able to construct creative objects given a set of inputs, whatever and however many there are (that can be represented using numbers) can indeed be tested for compatibility with the DSNS approach; but this is clearly beyond the scope of the present work.

Chapter 6
Conclusions

Abstract We conclude with an overview of the concept of creativity and how it must be a process, like any other, that can be mechanized. The advantages of being able to do so are clear as advancements in virtually every field should follow, unhindered by the slow and often clouded thinking of humans. The DSNS approach proposed is just such a process that is, in principle, both scalable and applicable in other domains. However, much more testing is required before any grandoise claims of its usefulness can be made, despite whatever has already been demonstrated in the domain of chess problem or puzzle composition. In conjuction with other technologies, such as robotics, the DSNS could also be a real-time component in sentient or semi-sentient androids, enabling them to come up with new ideas and suggestions that could benefit humans. Given the scale of problems facing humanity today (e.g. environment, medicine), computational creativity may be our only hope of survival into the distant future.

Keywords Conclusions · DSNS · Future · Robotics · Problems · Mechanization · Process

Human creativity has proven to be quite an elusive concept to formalize; more so than 'mere' intelligence. Our creativity is not only a source of enjoyment in terms of being able to create beautiful objects but it is also the source of all our technological advancements such as useful ideas and designs that have materialized, be it vaccines to prevent disease or computerized vehicles to take us into space. Essentially, everything around us of value is the product of human creativity. So the incentive to 'mechanize' this process—perhaps in a way that could far exceed our own abilities—is clear. To date there have been many attempts by computer scientists and artificial intelligence researchers to develop software that can 'come up' with creative products such as music, poetry and paintings. Some more successful than others; yet improvements over time cannot be denied.

However, to our knowledge, there is no single underlying process pertaining to computational creativity that is domain-independent. In this book, we have proposed just such a thing: the digital synaptic neural substrate (DSNS), named for its analogous nature to the generic, neuro-chemical substances that likely represent

A. Iqbal et al., *The Digital Synaptic Neural Substrate*,
SpringerBriefs in Cognitive Computation, DOI 10.1007/978-3-319-28079-0_6

objects and our experiences in our brains. That generic substance that mingles and interacts in poorly-understood ways so as to sometimes, in some people, create that 'Eureka!' moment. The DSNS has been shown in this research to be effective in the process of composing chess problems of reasonably high quality[1]—though not quite yet on the level of the *best* human composers—using information obtained from the same domain and *also* using information taken from the same domain *and* a completely unrelated domain. The reason for this is still an open question. A significant advantage of the DSNS approach is the lack of the need to program domain-specific, rule-based heuristics for creating creative content. It can be seen as an underlying process or backbone approach that can be 'plugged in' to any creative content generation system or tool by supplying the necessary attributes that system needs to do its work. The DSNS applies the necessary 'constraints' to the system to hone its output.

There are virtually an infinite number of different ways or permutations to further test the DSNS approach. Variations might include in terms of the domain type, domain subtypes, desired output type, length of random search for a matching deviation value, and even combining more than two domains. Not to mention variations in the number of attributes and objects used. Which combinations work best and why? So the DSNS approach appears to be quite scalable and open to many further inquiries for interested researchers. However, these tests are beyond the scope of this book which aims primarily to introduce the DSNS approach and concept for use in the field of computational creativity and perhaps even other fields.

From an artificial intelligence standpoint, one might imagine the DSNS approach implemented in an advanced robot capable of fully exploring its environment, using a variety of senses; perhaps even more and more sensitive than our own. A robot that is able to automatically extract attributes or features from the objects it perceives and 'pool' them together in the form of DSNS strings in its computerized brain. These strings will then be free to mingle until something of value—in any domain the robot is familiar with—can likely be produced. The robot would have a built-in system of trial and error that learns automatically which combinations of attributes from which domains produce the best results and would create a cycle of consistently improving output. This process would be, in principle, inexhaustible because the same creative objects produced can then be fed back into the robot's perception for further processing, assuming an ever-changing environment was not enough. A reality in the not-so-distant future, perhaps. In any case, at the very least, the DSNS approach has been shown to be quite good at composing valid chess problems of the three-move variety (that requires creativity). This is what we set out to do and have hopefully demonstrated in this book.

---

[1]This is with reference to 'traditional' chess problems that must abide by various composition conventions; not all of which have aesthetic merit. However, there is a new, comparable class of chess problems known as 'chess constructs' which have aesthetic value yet need not abide by such conventions. These are therefore less esoteric and more accessible to a wider audience of players and composers. By this standard, the compositions generated are likely of even higher quality.

Appendix A
Chess Problem Composing Steps

Chess problems for this research were composed automatically using CHESTHETICA (see Fig. A.1) following the essential steps below. It is a modification of the approach used in (Iqbal 2011). The problems composed are limited to orthodox mate-in-3 problems in standard international chess.

1. Obtain the two *new* DSNS strings produced from the DSNS process (see Sect. 3.1).
2. Set the total number of white pieces and black pieces that can be used in the composition. There are two attribute values pertaining to these features in each DSNS string.
   (a) For example, if in string 1 the white piece count is 5 and in string 2 the white piece count is 10, the range possible for this computer composition is between 5 and 10. The same for the black pieces.
3. Calculate the total Shannon value of the white pieces and then the black pieces in both strings and get the average of each. Use these average values to determine the number of piece permutations (i.e. combinations of different pieces) that satisfy them.
   (a) For example, an average value of 10 for white could mean having a bishop, two knights and a pawn whereas an average value of 9 for black could mean having just a queen. The total number of piece permutations possible for both the white and black pieces here is totaled.
   (b) This total will equal the number of times the same pair of DSNS strings is used in attempting to generate a composition. So, in principle, every legal piece combination can be tested.
4. Determine which permutations of pieces for both white and black satisfy (2) and are 'reasonable', i.e. in total, there are no more than 8 pieces—other than pawns—on the board.
   (a) For example, if the range of pieces for white that can be used in the composition is between 4 and 6, then permutations with only 3 pieces are excluded.

A. Iqbal et al., *The Digital Synaptic Neural Substrate*,
SpringerBriefs in Cognitive Computation, DOI 10.1007/978-3-319-28079-0

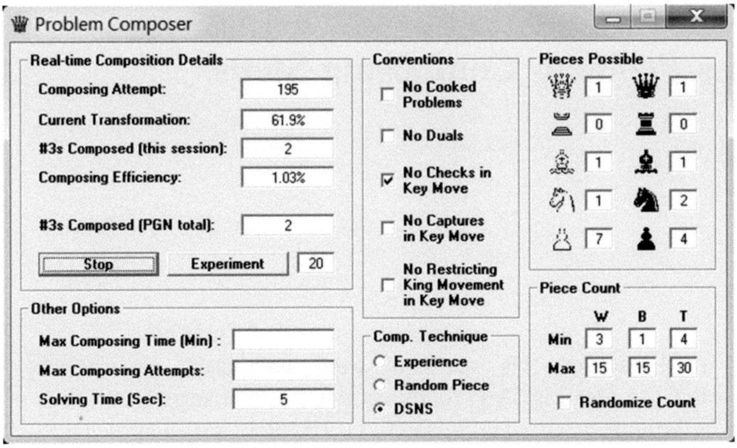

Fig. A.1 CHESTHETICA v9.22 composition interface window

(b) 'Reasonable' means that the position should be realistic. Typically, only one or two of the pieces on the board for a particular color would have a pawn promoted to it. So, if the upper limit of the total number of white pieces allowed (as per (2)) is 12 and you have a possible permutation with one queen, four rooks, four bishops and a knight, this will be excluded because even though the total number of pieces is 10 (below 12), it is more than 8 pieces. On the other hand, one queen, three rooks, two bishops and two knights would be acceptable and more realistic.

5. If no permutations for both white and black can be found that satisfy the requirements in (4), return to (1), otherwise use a random, valid one for each.
6. Place the two kings on random squares on the board. Accept them so far as the resulting position is legal; otherwise, keep repeating the process.
7. Choose at random one of six possibilities (i.e. the five remaining piece types and a 'blank square') based on equal probability (i.e. 1 in 6 or 16.67 %). Alternate between white (first) and then black.

   (a) If a 'blank' is chosen (which could be for either white or black) return to (7) and choose for the opponent's color instead. So a blank means one color misses its 'turn' and could therefore have fewer pieces on the board in the end.

8. Choose a random square until one that is unoccupied is found.

   (a) This is where, if a piece was chosen in (7), it will be placed.

9. Determine, based on the two DSNS strings, which are the first and last pieces to have moved.

10. If the piece chosen in (7) is not the same as any of the piece types determined in (9) and none of the latter are already on the board, set a 50 % chance that the former will have to be chosen again.
11. Place the chosen piece on the square determined in (8).
12. Check the legality of the position in terms of chess rules, taking into account the constraints mentioned earlier.

    (a) For example, having a pawn occupying the eighth rank is illegal.
    (b) The possibility of castling was given a 'neutral' 50 % random probability of being legal, assuming a king and one of its rooks happen to be on the right squares. Determination of legality based on retrograde analysis was considered too complicated and unnecessary for the purposes of this research. 'Officially', in compositions, castling in the key move is legal unless it can be proven otherwise.
    (c) En passant captures, if plausible, default to illegal. En passant is considered legal only if it can be proven the last move by the opponent permitted it.

13. If the position is illegal, remove the chosen piece from the board and return to (7).
14. Determine if the material difference between white and black for the position at present exceeds the higher of the two Shannon material differences in the two DSNS strings.
15. If (14) is true then clear the board and start composing a new problem; return to (3).

    (a) The same DSNS strings are used but with a refreshed piece permutation array.

16. Sum the sparsity values from both DSNS strings.
17. If the total from (16) ≥1 (i.e. leaning toward a sparser position) then determine if the sparsity value of the present position is less than 0.25 (i.e. leaning toward density). If so, clear the board and start composing a new problem; return to (3).
18. If the total from (16) <1 (i.e. leaning toward a denser or crowded position) then determine if the sparsity value of the present position is more than 0.75 (i.e. leaning toward a sparser position). If so, clear the board and start composing a new problem; return to (3).
19. Keep the piece chosen in (7) on the board and return to (7) to choose a new piece for the opponent's army until all the constraints above have been satisfied.
20. Test with a mate-solver engine to determine if the tentatively acceptable position generated has a forced mate-in-3 solution to it. If not, remove the last chosen piece from the board; return to (7).

    (a) CHESTHETICA communicates with ChestUCI v5.2 for this purpose (5 s search limit).

21. If there is such a solution, the position is ‘optimized’ as shown in the code below. This makes the composition more economical in form.

    (a) FOR every square

```
IF not occupied by a king and not empty THEN
    Remove piece
    IF forced mate-in-3 can still be found THEN
        Proceed
    ELSE
        Return piece to its original location
        END IF
    END IF
NEXT
```

    (b) To be thorough, optimization is performed three times, starting from the upper left to the lower right of the board; white pieces first, then black, and then white again. Fewer passes proved to be insufficient in certain positions. Optimization generally increases the aesthetic quality of a composition by removing unnecessary or passive pieces, but not always.
    (c) Sometimes, optimization may not be possible.

22. Test for conformity to composition conventions specified, if any. If there are any conventions specified not satisfied, pieces are added by returning to (7).

    (a) In the case of the convention ‘*no restricting enemy king movement in key move*’, the piece chosen earlier is actually removed before returning to (7) to help avoid this problem the next time around.
    (b) The constraints mentioned earlier on need not be re-applied after the optimization and conformity to convention processes.

23. Accept the composition as valid, optimized and in conformity with specified conventions.

    (a) The composition is stored in a PGN file.

24. Clear the board and return to (5) if the total number of permutations in 3(b) is not yet reached, otherwise, return to (1).

    (a) So a particular pair of DSNS strings can possibly be used to generate more than one composition (or perhaps none).

Appendix B
Two Columns of DSNS Strings in a Spreadsheet

A. Iqbal et al., *The Digital Synaptic Neural Substrate*,
SpringerBriefs in Cognitive Computation, DOI 10.1007/978-3-319-28079-0

COMP #3 > 3.5—Set A.pgn	White pieces	Black pieces	Shannon white pieces	Shannon black pieces	Shannon difference	#moves	Year	First piece to move	Last piece to move	Sparsity	|d|	Deviation	Σ÷
8/1p2BN1K/4Qp2/n1R4p/3k2P1/P5n1/4P3/1r6 w - - 0 1	8	7	23	14	9	3	1908	2	2	0.454545455	123.015	124.801	21.786
5rk1/5qpn/8/3N4/3B4/1B6/1KP3R1/8 w - - 0 1	6	5	15	18	3	3	1851	4	2	0.314285714	179.017	183.768	24.751
5Q2/b2k1P2/1n1NNn2/1P1p4/6P1/8/8/7K w - - 0 1	7	5	18	10	8	3	1883	2	1	0.352941176	63.314	66.947	23.633
r4Qb1/p1p3Pk/P6p/7P/8/6K1/5R2/8 w - - 0 1	6	6	17	11	6	3	1930	4	2	0.352941176	31.004	32.305	21.301
b7/3NQ3/3N4/1n1k2n1/8/p2K4/8/8 w - - 0 1	4	5	15	10	5	3	1888	2	2	0.529411765	60.129	62.435	22.306
r1q1r3/8/k1p1p3/1pK5/7R/1P6/8/4QB2 w - - 0 1	5	7	18	22	4	3	1849	5	1	0.545454545	60.165	61.925	21.760
5bb1/5p1n/3p1pkP/3K1NpN/6P1/8/8/8 w - - 0 1	5	8	8	13	5	3	1908	2	1	0.220338983	95.258	103.269	28.011
8/1K3Q2/2P5/4k1Pn/1P6/2PBbPpn/8/3b2N1 w - - 0 1	9	6	20	13	7	3	1885	1	1	0.483870968	115.183	118.963	23.780
3b4/2p5/6p1/6k1/8/3NN3/1Q6/5K2 w - - 0 1	4	4	15	5	10	3	1993	2	2	0.571428571	179.279	193.410	34.131
8/4p3/7B/4p3/4k1P1/8/5K2/3R4 w - - 0 1	4	3	9	2	7	3	1845	3	4	0.777777778	230.522	252.478	41.956
1Q1b4/r1n1ppB1/r7/R1Nb2k1/4B1p1/6P1/5nK1/7R w - - 0 1	8	10	29	25	4	3	Null	5	4	0.36	48.107	54.732	24.625
8/8/1Kp1N3/8/2k1r3/1R2p3/B7/2bR4 w - - 0 1	5	5	16	10	6	3	1989	2	2	0.5	60.100	62.705	20.605
K7/b3p2k/2p1B3/4QPPp/1p6/4p3/1p6/5n2 w - - 0 1	5	9	14	12	2	3	1997	1	1	0.5	100.152	100.768	20.616

(continued)

COMP #3 > 3.5—Set A.pgn	White pieces	Black pieces	Shannon white pieces	Shannon black pieces	Shannon difference	#moves	Year	First piece to move	Last piece to move	Sparsity	|d|	Deviation	∑÷
8/1k2N3/RB5K/1Bp1N3/8/3p4/2P4p/1r5r w - - 0 1	7	6	18	13	5	3	1924	4	1	0.393939394	77.606	87.730	30.124
5n1k/8/8/5Q2/5q1B/8/8/6KR w - - 0 1	4	3	17	12	5	3	1935	5	5	0.636363636	30.136	34.832	24.696
8/1R6/8/1Q1n4/8/b4K1p/8/6k1 w - - 0 1	3	4	14	7	7	3	1996	5	4	1	48.625	54.372	23.747
B3R2n/4Pr2/1KNN4/1ppk1Ppb/3p1RP1/P1P5/4P3/Qq2b3 w - - 0 1	13	10	34	27	7	3	1973	2	3	0.291139241	112.288	119.379	27.091
1b1q4/1N1N4/2B5/4p3/1Pk5/7p/1K6/3RB3 w - - 0 1	7	5	18	14	4	3	1971	6	2	0.5	106.194	108.306	22.112
Kb1Q2b1/n1p5/1R1pr3/1B1k1P2/R1NPN2P/2B2r2/3n1P2/1q6 w - - 0 1	12	10	35	33	2	3	1972	5	3	0.297297297	21.026	22.298	21.273
8/8/6B1/6p1/8/6K1/4k3/2R2N2 w - - 0 1	4	2	11	1	10	3	Null	4	3	0.6	39.284	61.253	39.969
8/3PnknP/5N2/6K1/8/8/8/8 w - - 0 1	4	3	5	6	1	3	1957	1	2	0.304347826	51.096	52.746	21.650
1R6/2PR4/8/1N6/8/6Q1/8/kn3K2 w - - 0 1	6	2	23	3	20	3	1952	4	2	0.571428571	63.000	64.111	21.111
3b3r/1bn1P3/1r5Q/pPk1p3/K2N1P2/1B1P4/4P2p/6B1 w - - 0 1	10	9	23	22	1	3	1875	3	1	0.333333333	121.095	122.949	21.854
2n5/6K1/4p3/6np/8/8/5N1k/4BQ2 w - - 0 1	4	5	15	8	7	3	1914	6	2	0.529411765	106.257	110.677	24.420
1b2K3/rn1P1p2/pP1k1PpP/2Rbp1NQ/3Pp3/7P/2NR1q2/3r4 w - - 0 1	12	12	31	33	2	3	1968	2	1	0.244897959	149.103	151.354	22.251

(continued)

COMP #3 > 3.5—Set A.pgn	White pieces	Black pieces	Shannon white pieces	Shannon black pieces	Shannon difference	#moves	Year	First piece to move	Last piece to move	Sparsity	|d|	Deviation	Σ÷
1Q1b4/3Pr3/3R1B1n/1N2RpP1/K4kp1/4pP2/1Pp1Pp1P/n1rq4 w - - 0 1	12	12	31	33	2	3	1999	2	4	0.25	25.053	23.885	18.832
1N6/3P1Pb1/r2k2P1/q3R3/ppR1P3/5p1K/Q7/Bn6 w - - 0 1	10	8	29	23	6	3	1981	5	1	0.375	56.067	70.108	34.041
2B5/8/1N1P4/4kpP1/8/B4K2/4P3/8 w - - 0 1	7	2	12	1	11	3	1904	1	2	0.529411765	148.115	182.313	54.199
N7/2Q3Nb/2p5/n2p2p1/3B4/2Pk4/8/b4K2 w - - 0 1	6	7	19	12	7	3	1910	5	2	0.481481481	8.166	8.761	18.596
8/8/5Nb1/8/1Q1N1k2/8/7K/8 w - - 0 1	4	2	15	3	12	3	Null	6	2	0.75	30.446	37.231	24.786
3n4/6p1/3p4/4k3/1pp4Q/4N3/6K1/4R3 w - - 0 1	4	6	17	7	10	3	1941	5	2	0.714285714	80.078	85.006	24.928
6k1/5pP1/5K1P/8/B7/5N2/8/8 w - - 0 1	5	2	8	1	7	3	1976	3	1	0.368421053	90.053	114.418	44.365
8/1pR5/4p3/4k1K1/8/2Np2p1/3P2B1/8 w - - 0 1	5	5	12	4	8	3	Null	2	1	0.454545455	29.066	35.001	23.935
6bK/3B3p/1b1p4/2pkr1NR/R3p3/1P2p3/1nP1P2p/4Q2r w - - 0 1	9	12	28	25	3	3	1996	2	1	0.333333333	210.667	268.720	78.053
8/8/8/5N1k/8/5K1N/8/3R4 w - - 0 1	4	1	11	0	11	3	1983	2	4	1	100.632	104.838	24.206
n7/r1RP1K2/3p4/B1pk1pPQ/N2bpP2/n7/P1NrB3/3R2q1 w - - 0 1	12	11	35	32	3	3	1957	5	1	0.315068493	141.152	143.831	22.680
4RQ2/1rp5/PbP1NRKp/3k1p2/Pp3pP1/1n3b2/2nBPq2/2rN1B2 w - - 0 1	13	13	36	36	0	3	1955	5	2	0.25	89.329	91.504	22.175

(continued)

COMP #3 > 3.5—Set A.pgn	White pieces	Black pieces	Shannon white pieces	Shannon black pieces	Shannon difference	#moves	Year	First piece to move	Last piece to move	Sparsity	|d|	Deviation	∑÷
6br/3pP3/pK3Q1p/3k4/8/p7/1pPP1P2/qrn5 w - - 0 1	6	11	13	30	17	3	1918	1	1	0.361702128	61.044	69.378	28.334
8/1B6/2NB1P2/1p1k3P/2b1R2K/8/3P4/3Q4 w - - 0 1	9	3	26	4	22	3	1887	3	2	0.375	62.096	71.681	29.585
b7/4pP2/5pRp/R2q1k1B/4np1K/3P4/1Q3p2/8 w - - 0 1	7	9	24	20	4	3	1992	5	4	0.333333333	31.086	33.843	20.757
5q1n/R4P1k/4P1Nb/8/3P4/8/B1Q5/K6R w - - 0 1	9	4	28	15	13	3	1935	1	2	0.448275862	47.178	55.287	28.108
K6R/1PP1k1n1/6pP/6PP/1rN2N2/Q1p4b/8/1r6 w - - 0 1	10	7	25	18	7	3	1924	5	1	0.346938776	83.107	85.546	22.439
6Qb/1p1B4/1N3B2/2k2p2/1N3p1q/4p2p/PPb5/2Kn4 w - - 0 1	8	10	23	23	0	3	1966	2	1	0.391304348	151.109	162.077	30.969
5k1K/3p4/3N4/4P1N1/1B6/8/8/8 w - - 0 1	5	2	10	1	9	3	1904	3	2	0.636363636	52.136	60.184	28.048
5B2/5P1P/4P1k1/8/6K1/8/8/8 w - - 0 1	5	1	6	0	6	3	1994	1	1	0.428571429	110.117	123.907	33.790
Q3B3/6N1/2NpRp1n/2Pk2n1/3P2P1/1Kp4p/1b6/7b w - - 0 1	9	9	26	16	10	3	Null	5	2	0.3	40.025	49.615	27.590
2n5/2K2p2/p7/p1pk1n2/2R4Q/P2b1p1N/1B3N2/8 w - - 0 1	7	9	24	14	10	3	1881	2	1	0.470588235	139.183	140.642	21.459
8/6K1/pN1Q4/1k2p3/3bN1R1/8/nPP5/7B w - - 0 1	8	5	25	8	17	3	1912	2	2	0.419354839	61.008	70.005	28.998

(continued)

COMP #3 > 3.5—Set A.pgn	White pieces	Black pieces	Shannon white pieces	Shannon black pieces	Shannon difference	#moves	Year	First piece to move	Last piece to move	Sparsity	|d|	Deviation	∑÷
k7/1r6/1P4b1/8/8/1R4K1/8/7B w - - 0 1	4	3	9	8	1	3	1965	4	4	0.636363636	67.265	72.229	24.964
q2Q4/2p5/1bk1n3/4PB2/N1n2PK1/3p4/2N5/8 w - - 0 1	7	7	20	20	0	3	1880	5	2	0.35	171.087	170.593	19.506
8/3K1PBk/8/7P/8/4N3/8/8 w - - 0 1	5	1	8	0	8	3	1992	1	2	0.6	105.322	114.792	29.470
q4B2/4P1n1/4k3/1b2NrP1/1Q3P2/4K3/8/8 w - - 0 1	7	5	18	20	2	3	1855	5	2	0.333333333	105.039	110.070	25.031
5r2/p2R1P1k/5p1p/8/q7/1p6/PP4QP/K7 w - - 0 1	7	7	18	18	0	3	Null	5	4	0.411764706	**19.062**	**18.217**	**17.155**
1K3RBb/n3P2p/1Np1PPp1/1pP1k3/4p2P/2PpB1N1/b4r2/nQ6 w - - 0 1	13	12	32	23	9	3	1984	1	2	0.287356322	63.141	69.504	26.362
3k3n/p1R5/8/1B2B3/4K3/8/8/8 w - - 0 1	4	3	11	4	7	3	1911	4	4	0.636363636	47.011	49.615	22.603
8/5N2/2p5/8/3RPk2/NB5P/2P2K2/8 w - - 0 1	8	2	17	1	16	3	1954	4	2	0.555555556	**34.056**	**49.667**	**33.612**
6B1/4pPq1/5k2/8/5P1r/8/8/1QK1R3 w - - 0 1	6	4	19	15	4	3	1994	4	5	0.384615385	58.027	62.711	24.684

COMP #3 > 3.5 - Set B.pgn	White pieces	Black pieces	Shannon white pieces	Shannon black pieces	Shannon difference	#moves	Year	First piece to move	Last piece to move	Sparsity
8/8/4p3/2N1K3/RP2P3/2k5/p1P1Q3/b7 w - - 0 1	7	4	20	5	15	3	2009	2	2	0.44
1n1n3q/R1prBN2/1Nk1p3/Qp5r/2p2P1P/2P5/3P2R1/5K1B w - - 0 1	12	10	35	29	6	3	1982	1	2	0.297297297
1k6/1P6/P7/2P1P3/8/Q7/8/K4b2 w - - 0 1	6	2	13	3	10	3	1927	1	1	0.666666667

(continued)

COMP #3 > 3.5 - Set B.pgn	White pieces	Black pieces	Shannon white pieces	Shannon black pieces	Shannon difference	#moves	Year	First piece to move	Last piece to move	Sparsity
1K4N1/5np1/4N2b/3ppB1k/8/5pP1/Q3p2P/8 w - - 0 1	7	8	20	11	9	3	1949	2	1	0.348837209
8/1K6/8/kp6/1n6/1PB1p3/8/1N6 w - - 0 1	4	4	7	5	2	3	1930	3	2	0.4
8/7r/1p1pN1Q1/np2k3/3R3P/n4p2/KPP5/3b4 w - - 0 1	7	9	20	18	2	3	1894	2	1	0.380952381
8/1P3P1R/1P1k1P2/2p5/2p1K2P/2P5/8/8 w - - 0 1	8	3	11	2	9	3	1975	1	2	0.47826087
3N2R1/r1pN1B2/4pp1r/1P1kqb1R/1Q6/3pnpBp/1b2PP2/n6K w - - 0 1	11	14	34	37	3	3	1947	2	1	0.301204819
1B6/1n2K3/1rprNN2/4k3/1p1pbpP1/R7/2nP1P2/2R5 w - - 0 1	9	10	22	23	1	3	1862	4	1	0.292307692
K7/4QP1B/2pp1rR1/1PPk1r1p/P2p1P1q/BP2bbP1/1N6/4R3 w - - 0 1	14	10	35	29	6	3	2000	2	1	0.255319149
k2r4/1b6/N7/1PK1Q3/8/8/8/8 w - - 0 1	4	3	13	8	5	3	1814	5	1	0.466666667
8/8/B7/1Np4Q/4k3/6K1/7N/1n2n3 w - - 0 1	5	4	18	7	11	3	1940	2	2	0.6
8/3p3Q/2rp1q1B/R1nN2N1/3k4/1P3p1r/K3P3/8 w - - 0 1	8	8	25	25	0	3	1929	3	1	0.347826087
4n3/P7/4K1R1/8/8/8/2R5/5k2 w - - 0 1	4	2	11	3	8	3	1968	1	4	1
4n3/1p2Kp2/8/4k2p/2p1PR2/p7/5P1p/bQ6 w - - 0 1	5	9	16	12	4	3	1952	5	1	0.5
B3b2K/1n2Rp2/2Pk4/5Pp1/1P3N2/3pp1B1/1R5b/Q7 w - - 0 1	10	8	31	13	18	3	Null	4	2	0.375
3N1B2/8/qP2NkBK/8/3P1P2/7P/8/8 w - - 0 1	9	2	16	9	7	3	1911	2	1	0.578947368
8/8/4Rp1r/1Q1p2Bp/nK1kp3/1p3r2/3nP3/8 w - - 0 1	5	10	18	21	3	3	1884	3	1	0.306122449
rQ4q1/2R5/3N1pRp/r2p3n/1NpP1kPB/5P2/5PK1/2n5 w - - 0 1	11	10	32	29	3	3	1964	2	2	0.323076923

(continued)

COMP #3 > 3.5 - Set B.pgn	White pieces	Black pieces	Shannon white pieces	Shannon black pieces	Shannon difference	#moves	Year	First piece to move	Last piece to move	Sparsity
3n4/K1k1p2n/1Nb2B1r/1P1QP3/8/8/8/8 w - - 0 1	6	6	17	15	2	3	1872	1	1	0.315789474
3nK3/8/1b4B1/1p2k3/2pp1R2/3p1p2/7B/8 w - - 0 1	4	8	11	11	0	3	1926	3	3	0.4
8/1N4Q1/n7/1N1k4/4b3/K7/8/3B4 w - - 0 1	5	3	18	6	12	3	1908	3	2	0.571428571
b1QN1b2/5p1r/3p1B2/1p1k1KNp/2p5/n1p2P2/2pp4/r5R1 w - - 0 1	7	14	24	27	3	3	1979	4	1	0.428571429
1rk1bN2/Pp1Rp1Pp/1B1pPp1b/8/KP6/8/R1B4r/1Q6 w - - 0 1	11	10	32	21	11	3	1972	5	1	0.272727273
8/1b5r/1pNp2p1/4pp2/1P2k1BQ/2K1N3/2n3P1/8 w - - 0 1	7	9	20	16	4	3	1858	2	2	0.347826087
N7/KPB5/2P1r3/2Q2p2/1RNPk3/1p4P1/bb2P1p1/3r2q1 w - - 0 1	11	9	28	28	0	3	2007	3	2	0.303030303
8/8/8/8/5NQ1/1Np5/1pR2K2/1k6 w - - 0 1	5	3	20	2	18	3	1981	2	2	0.307692308
1b6/5N2/q2p3p/7p/1nprBk1K/r2p4/3P1N2/6R1 w - - 0 1	6	11	15	30	15	3	2003	4	2	0.414634146
4r2B/2N5/p7/7r/3p4/2k3N1/1R6/K4B2 w - - 0 1	6	5	17	12	5	3	Null	3	2	0.647058824
8/4Q3/8/8/1p2pNPb/1Np1Rb2/K1n5/3k1B2 w - - 0 1	7	7	24	12	12	3	1958	2	2	0.304347826
7r/1k2PP2/3Q4/1N6/8/8/6K1/8 w - - 0 1	5	2	14	5	9	3	2005	1	3	0.636363636
n7/RPNkp2b/7b/r1p2p2/1p6/3pRK2/1Q6/8 w - - 0 1	6	10	23	19	4	3	1936	5	4	0.421052632
1b1nN2B/7Q/3p4/6k1/6p1/2P2prp/6b1/3K4 w - - 0 1	5	9	16	18	2	3	1961	2	2	0.388888889
8/7P/8/8/8/5K1p/8/7k w - - 0 1	2	2	1	1	0	3	1861	1	4	1
5Nkb/PPpp2n1/3P4/3P3B/2P2K2/8/6R1/8 w - - 0 1	9	5	16	8	8	3	1912	5	3	0.368421053

(continued)

COMP #3 > 3.5 - Set B.pgn	White pieces	Black pieces	Shannon white pieces	Shannon black pieces	Shannon difference	#moves	Year	First piece to move	Last piece to move	Sparsity
8/2b4r/K1p1pN2/4k1pp/8/2PR4/B4Q2/5n2 w - - 0 1	6	8	21	15	6	3	1861	4	2	0.466666667
qb5K/2n1R2P/7k/2n5/6Q1/8/8/5b1r w - - 0 1	4	7	15	26	11	3	1925	4	1	0.578947368
7Q/2q5/n7/4Nn2/r2R4/1Rp1kB2/p1b5/ K3B3 w - - 0 1	7	8	28	25	3	3	1936	5	2	0.405405405
5r2/4n1p1/1q1RNb2/1p2k3/1Q3NK1/4r1pn/ p4p2/8 w - - 0 1	5	12	20	33	13	3	1883	2	2	0.278688525
8/8/Kb1B2B1/5R2/2Ppk1P1/2p5/4pp2/2R5 w - - 0 1	7	6	18	7	11	3	Null	4	3	0.419354839
8/6p1/6Q1/1NnNpp2/1pB1k3/8/p1b1PbK1/ 5rn1 w - - 0 1	6	11	19	22	3	3	1941	5	1	0.26984127
3r4/2p1pPK1/1qPpkp2/Rp1N1p2/4Np1P/ 5P2/3RB1n1/Br3Q2 w - - 0 1	12	12	35	29	6	3	1974	3	3	0.24
B7/3K4/8/3Q1p2/8/1pk5/1N6/B7 w - - 0 1	5	3	18	2	16	3	1871	5	2	0.5
5B2/4N1K1/2B5/1Pp1kb2/8/b1P1p3/8/3Q4 w - - 0 1	7	5	20	8	12	3	1879	5	2	0.5
6N1/4p1RP/4kp2/KN1r2rp/2Q2B2/ p1p1PPn1/q7/8 w - - 0 1	9	10	26	27	1	3	1950	1	2	0.31147541
1n4B1/2p2q1r/p1NPrp1p/K1B5/Q2Nk1Pp/ 1RP3P1/p3PP2/7b w - - 0 1	13	12	32	31	1	3	1994	3	1	0.324675325
6n1/1n2q2p/2QR1p2/r1PPp3/1p1Nk3/5R2/ P4P1P/B4K1B w - - 0 1	12	9	33	24	9	3	1994	1	1	0.287671233
5b2/1N6/rpp3K1/5Q2/k1B1n3/p7/N1Pn4/8 w - - 0 1	6	8	19	17	2	3	1934	5	1	0.411764706
8/8/6Qb/4k2r/8/nKN1p2P/2P1N1Pr/8 w - - 0 1	7	6	18	17	1	3	1927	2	1	0.371428571
2nr4/q1R5/1PN1N2Q/PBPk1P2/1p1r3B/ 5PP1/b1p1Pn2/b3RK2 w - - 0 1	15	10	38	33	5	3	2004	5	2	0.263157895

(continued)

COMP #3 > 3.5 - Set B.pgn	White pieces	Black pieces	Shannon white pieces	Shannon black pieces	Shannon difference	#moves	Year	First piece to move	Last piece to move	Sparsity
3Q1br1/1p1rPP1b/3P2p1/N1p3P1/3kpK2/2Np4/1P1n4/8 w - - 0 1	9	11	20	24	4	3	1941	1	2	0.277777778
K1N2R2/6p1/2B1kb1p/1p6/p4P2/1p1pP2P/5Pp1/6Qb w - - 0 1	9	10	24	13	11	3	1930	5	1	0.37254902
6k1/1p1K2n1/8/6N1/rb3QN1/8/8/8 w - - 0 1	4	5	15	12	3	3	1889	5	2	0.473684211
QK5k/6Rp/3nb3/5p2/1p3p2/p7/8/BqR1r3 w - - 0 1	5	10	22	25	3	3	1955	5	4	0.428571429
1R6/3P4/8/2Pk1PNB/8/B1K5/8/5N2 w - - 0 1	9	1	20	0	20	3	1901	2	2	0.625
1n6/3q2B1/3R3Q/2k1N3/6p1/1nN3K1/4P3/8 w - - 0 1	7	5	24	16	8	3	Null	4	2	0.5
4Nk2/3P1P2/1N4P1/8/6K1/8/8/8 w - - 0 1	6	1	9	0	9	3	1975	1	2	0.411764706

Appendix C
The 90 DSNS-Generated Compositions Evaluated by the Human Experts

#	FEN	Solution	Source
1	3K4/3b4/4p2k/8/8/4R3/8/1R6 w - - 0 1	1. Rg1 e5 2. Kxd7 e4 3. Rh3# 0-1	tg2500+photo
2	1b6/P7/k4r2/8/1P6/1KP5/2R1P3/8 w - - 0 1	1. axb8=Q Rb6 2. Qa8+ Kb5 3. c4# 0-1	tg2500+photo
3	2k5/7P/8/8/K7/b7/1p3Q2/8 w - - 0 1	1. Qf7 b1=Q 2. h8=Q+ Bf8 3. Qhxf8# 0-1	tg1500+photo
4	8/8/8/4Q3/2k5/1R3q2/8/1K6 w - - 0 1	1. Rxf3 Kb4 2. Rf4+ Kb3 3. Qe3# 0-1	tg2500+photo
5	8/6b1/P5R1/2B5/3R4/7K/4n3/6bk w - - 0 1	1. a7 Nf4+ 2. Rxf4 Bh2 3. a8=Q# 0-1	tg2500+photo
6	8/8/8/R1K5/1r6/8/1k6 w - - 0 1	1. Kxb3 Kc1 2. Rd4 Kb1 3. Rd1# 0-1	tg1500+photo
7	6B1/8/6KR/8/4k3/8/7Q/8 w - - 0 1	1. Qd2 Ke5 2. Rh5+ Ke4 3. Bd5# 0-1	tg2500+photo
8	1B6/3kP3/3P4/1P1p4/5R1K/5b2/8/8 w - - 0 1	1. Rf8 Bh5 2. Kxh5 d4 3. e8=Q# 0-1	tg2500+photo
9	Q7/8/3Q4/8/1K2n3/8/5k2/8 w - - 0 1	1. Qxe4 Kf1 2. Qd1+ Kf2 3. Qde1# 0-1	tg1500+photo
10	8/6B1/3n4/5B1K/3R4/pp6/k7/8 w - - 0 1	1. Rd1 b2 2. Be6+ Nc4 3. Bxc4# 0-1	tg1500+photo
11	8/1K6/8/8/8/k1N5/6R1/2R5 w - - 0 1	1. Nd1 Kb3 2. Rb2+ Ka3 3. Ra1# 0-1	comp3.5
12	8/k2P4/8/3N4/8/1n4K1/8/8 w - - 0 1	1. d8=Q Na5 2. Qb6+ Ka8 3. Nc7# 0-1	tg2500+photo
13	8/1Q6/7K/3B4/8/6k1/8/8 w - - 0 1	1. Qf7 Kh3 2. Qf2 Kg4 3. Be6# 0-1	comp3.5
14	kN6/8/K7/4N3/8/8/3R4/b7 w - - 0 1	1. Rd8 Bxe5 2. Nc6+ Bb8 3. Rxb8# 0-1	tg1500+photo
15	4k3/bQ6/8/6K1/8/6R1/8/8 w - - 0 1	1. Qxa7 Kf8 2. Re3 Kg8 3. Re8# 0-1	tg1500+photo
16	8/8/8/4K3/1Q5n/8/k7 w - - 0 1	1. Kd3 Nf4+ 2. Kc2 Nd5 3. Qb1# 0-1	comp3.5
17	8/4K3/5Nk1/6B1/6r1/8/8/6R1 w - - 0 1	1. Rxg4 Kg7 2. Bf4+ Kh8 3. Rg8# 0-1	tg1500+photo

(continued)

A. Iqbal et al., *The Digital Synaptic Neural Substrate*, SpringerBriefs in Cognitive Computation, DOI 10.1007/978-3-319-28079-0

#	FEN	Solution	Source
18	1Rn5/8/4K3/8/8/kp6/3B4/2R5 w - - 0 1	1. Bc3 b2 2. Bxb2+ Ka4 3. Ra1# 0-1	comp3.5
19	8/P3k1P1/1p3R1p/8/p4K2/8/4p3/8 w - - 0 1	1. g8=Q h5 2. Qg7+ Ke8 3. a8=Q# 0-1	tg1500+photo
20	1k6/1P1R4/4p3/1K6/8/4n3/8/8 w - - 0 1	1. Kc6 e5 2. Rd8+ Ka7 3. Ra8# 0-1	tg2500+photo
21	7B/2K4P/8/7k/8/8/6R1/7n w - - 0 1	1. Bf6 Ng3 2. Rxg3 Kh6 3. h8=Q# 0-1	tg1500+photo
22	5Nk1/8/3R4/8/8/8/K7/2R5 w - - 0 1	1. Rd7 Kh8 2. Ng6+ Kg8 3. Rc8# 0-1	tg1500+photo
23	8/8/3K4/7B/8/4p2k/8/1Q6 w - - 0 1	1. Qg1 e2 2. Bxe2 Kh4 3. Qg4# 0-1	tg2500+photo
24	8/8/K1B2Q2/8/1k6/8/5b2/2N5 w - - 0 1	1. Bb5 Bg3 2. Nd3+ Kb3 3. Qb2# 0-1	comp3.5
25	5n2/4K1Pb/1R6/8/8/k7/2p5/8 w - - 0 1	1. gxf8=Q Be4 2. Qf1 c1=Q 3. Qa6# 0-1	comp3.5
26	k7/8/1P6/P2R4/2R5/3rN3/8/6K1 w - - 0 1	1. a6 Rd1+ 2. Rxd1 Kb8 3. Rd8# 0-1	comp3.5
27	8/7R/2p5/5K2/8/7p/1p1R4/7k w - - 0 1	1. Rb7 h2 2. Rbxb2 c5 3. Rb1# 0-1	comp3.5
28	5k2/2P5/8/8/5K2/8/8/3Rn3 w - - 0 1	1. Rd7 Nd3+ 2. Kf5 Nb4 3. c8=Q# 0-1	tg1500+photo
29	2k5/K3Pb2/2P5/8/p1r2b2/8/5Q2/4R3 w - - 0 1	1. Qb6 Be3 2. Rxe3 a3 3. Qd8# 0-1	tg1500+photo
30	8/1PK5/k5q1/4N3/bP6/8/3N4/8 w - - 0 1	1. Nxg6 Bb5 2. b8=Q Bc6 3. Qb6# 0-1	tg2500+photo
31	8/2R5/1P6/6K1/1k1p4/6p1/8/8 w - - 0 1	1. b7 g2 2. b8=Q+ Ka4 3. Ra7# 0-1	comp3.5
32	4k1B1/8/2KB4/6N1/6n1/8/8/8 w - - 0 1	1. Kc7 Ne5 2. Ne6 Nc6 3. Ng7# 0-1	comp3.5
33	8/3P2kr/8/K1R4p/8/8/8/4R3 w - - 0 1	1. d8=Q h4 2. Qd7+ Kg8 3. Re8# 0-1	tg1500+photo
34	6kN/4R3/4K3/8/8/8/8/8 w - - 0 1	1. Kf6 Kxh8 2. Kg6 Kg8 3. Re8# 0-1	tg1500+photo
35	1K3k2/1P6/4PNBP/8/8/4N3/r7/7n w - - 0 1	1. Nf5 Ra8+ 2. bxa8=Q Nf2 3. e7# 0-1	comp3.5
36	8/8/1R6/4k3/2K5/8/2QP4/8 w - - 0 1	1. Qg6 Kf4 2. Rf6+ Ke5 3. d4# 0-1	tg2500+photo
37	8/3P4/2k1K3/p7/1R6/2P5/8/8 w - - 0 1	1. Rb8 a4 2. d8=Q a3 3. Qd6# 0-1	comp3.5
38	8/2B3P1/8/7k/8/8/7K/8 w - - 0 1	1. g8=Q Kh6 2. Bf4+ Kh5 3. Qg5# 0-1	tg1500+photo
39	8/1R2K3/6bR/8/k7/8/8/8 w - - 0 1	1. Rxg6 Ka5 2. Rh6 Ka4 3. Ra6# 0-1	comp3.5
40	4K3/5R2/8/4b2k/6R1/8/3B4/8 w - - 0 1	1. Rg8 Bg7 2. Rfxg7 Kh4 3. Rh8# 0-1	tg2500+photo
41	8/1R6/1q6/8/8/8/K1R5/3N3k w - - 0 1	1. Rxb6 Kg1 2. Nc3 Kh1 3. Rb1# 0-1	tg2500+photo

(continued)

#	FEN	Solution	Source
42	8/2P5/8/8/8/8/k1K5/8 w - - 0 1	1. c8=Q Ka3 2. Qg4 Ka2 3. Qa4# 0-1	tg2500+photo
43	8/8/1B5R/4K2B/8/8/8/4k3 w - - 0 1	1. Be3 Kf1 2. Bf3 Ke1 3. Rh1# 0-1	comp3.5
44	7R/1p6/p7/8/1b2R2p/5K2/5p2/5k2 w - - 0 1	1. Rhxh4 Kg1 2. Reg4+ Kf1 3. Rh1# 0-1	tg1500+photo
45	7k/8/8/3B2K1/4p3/8/8/6B1 w - - 0 1	1. Kg6 e3 2. Bh2 e2 3. Be5# 0-1	tg1500+photo
46	B3k3/8/4KP2/4B3/8/8/8/8 w - - 0 1	1. Bd6 Kd8 2. f7 Kc8 3. f8=Q# 0-1	tg2500+photo
47	1K6/8/8/8/1k2N3/8/Q7/1B6 w - - 0 1	1. Bc2 Kb5 2. Qa7 Kb4 3. Qa4# 0-1	tg2500+photo
48	8/4P2n/2K5/8/7k/5R2/6R1/8 w - - 0 1	1. e8=Q Nf8 2. Qd8+ Kh5 3. Qg5# 0-1	comp3.5
49	8/R4P2/4k3/1B6/6p1/4K3/8/8 w - - 0 1	1. f8=Q g3 2. Bc4+ Ke5 3. Qf4# 0-1	tg1500+photo
50	8/7P/8/4pR2/8/p6K/8/6k1 w - - 0 1	1. h8=Q a2 2. Qg8+ Kh1 3. Qg2# 0-1	tg1500+photo
51	8/K2N3P/8/6k1/3N2P1/5nPP/1n6/8 w - - 0 1	1. h8=Q Kg6 2. Qg8+ Kh6 3. Nf5# 0-1	comp3.5
52	8/4BP2/2N5/3k4/4p3/2K5/8/8 w - - 0 1	1. Nd4 e3 2. f8=Q e2 3. Qf5# 0-1	tg2500+photo
53	8/8/6R1/3k4/8/8/RR6/2K5 w - - 0 1	1. Rgb6 Ke5 2. R2b4 Kf5 3. Ra5# 0-1	tg2500+photo
54	2R5/3P4/8/1k6/8/1p1K4/8/R5b1 w - - 0 1	1. d8=Q Bb6 2. Qe8+ Kb4 3. Qa4# 0-1	comp3.5
55	8/8/3Qr3/kB1n4/8/N7/1K6/8 w - - 0 1	1. Qc5 Re2+ 2. Bxe2+ Ka4 3. Bd1# 0-1	tg2500+photo
56	1n6/P1K5/5Q2/8/7B/8/4k3/8 w - - 0 1	1. axb8=Q Kd2 2. Qb3 Ke2 3. Qf2# 0-1	tg2500+photo
57	8/8/6p1/7k/8/8/2Q3n1/2BK3N w - - 0 1	1. Qxg2 Kh4 2. Qg3+ Kh5 3. Qh3# 0-1	comp3.5
58	8/6Q1/8/8/8/7B/1K6/4k3 w - - 0 1	1. Qg2 Kd1 2. Bg4+ Ke1 3. Qe2# 0-1	tg2500+photo
59	3K4/8/3k4/6R1/4R2b/8/8/8 w - - 0 1	1. Rxh4 Ke6 2. Rf4 Kd6 3. Rf6# 0-1	tg2500+photo
60	8/5P1k/1Rr5/4R3/2P5/3p4/8/5K2 w - - 0 1	1. f8=Q Rf6+ 2. Rxf6 d2 3. Re7# 0-1	tg1500+photo
61	5Nk1/1p6/8/8/8/8/8/K3Q3 w - - 0 1	1. Qe7 b6 2. Ne6 b5 3. Qg7# 0-1	tg2500+photo
62	8/6Q1/8/1K1k4/8/8/8/6N1 w - - 0 1	1. Qf6 Ke4 2. Kc4 Ke3 3. Qd4# 0-1	tg1500+photo
63	5N2/6R1/5k1N/3Pp3/8/8/1K6/8 w - - 0 1	1. Ne6 e4 2. Ng4+ Kf5 3. Rg5# 0-1	tg1500+photo
64	1k6/4p3/8/8/RK6/8/8/6BB w - - 0 1	1. Bb6 Kc8 2. Bc6 e6 3. Ra8# 0-1	comp3.5
65	4K3/1P4p1/R4b2/2bk4/5RN1/8/8/8 w - - 0 1	1. b8=Q Bce7 2. Qa8+ Kc5 3. Qc6# 0-1	tg1500+photo

(continued)

#	FEN	Solution	Source
66	8/8/8/2p1p3/8/k7/2B5/1K2R3 w - - 0 1	1. Re4 c4 2. Rxc4 e4 3. Ra4# 0-1	comp3.5
67	8/4N3/1Q6/1b6/8/2K2B1k/8/8 w - - 0 1	1. Nf5 Bc6 2. Qb8 Bd7 3. Qg3# 0-1	comp3.5
68	k7/2P5/3K4/8/8/8/8/8 w - - 0 1	1. Kc6 Ka7 2. c8=R Ka6 3. Ra8# 0-1	comp3.5
69	8/8/8/8/k4nK1/2P5/2P3Q1/8 w - - 0 1	1. Qb7 Ka5 2. Kxf4 Ka4 3. Qb4# 0-1	tg2500+photo
70	5K1k/8/1B6/8/8/1p6/4n3/2R3N1 w - - 0 1	1. Rc7 b2 2. Nxe2 b1=Q 3. Bd4# 0-1	tg1500+photo
71	8/3P1N2/3NN3/3k1p1p/3B4/8/8/2K5 w - - 0 1	1. d8=Q h4 2. Qc8 h3 3. Qc4# 0-1	comp3.5
72	4N3/3K4/4N3/4k3/4B3/8/2p2P2/3N4 w - - 0 1	1. Bxc2 Kd5 2. Ne3+ Ke5 3. f4# 0-1	tg1500+photo
73	2k5/8/4P3/4B3/4R3/6K1/8/8 w - - 0 1	1. Rb4 Kd8 2. Bd6 Ke8 3. Rb8# 0-1	tg1500+photo
74	8/3K2R1/8/3Q4/8/8/5k2/2n5 w - - 0 1	1. Qe4 Na2 2. Rg2+ Kf1 3. Qe2# 0-1	comp3.5
75	8/3N4/6Q1/8/2k5/B5K1/8/8 w - - 0 1	1. Nc5 Kd4 2. Qd6+ Kc4 3. Qd3# 0-1	comp3.5
76	8/8/8/5r1Q/8/1k6/8/R1K5 w - - 0 1	1. Qxf5 Kc3 2. Ra4 Kb3 3. Qc2# 0-1	tg2500+photo
77	8/8/3Rr3/8/1R6/5k2/8/7K w - - 0 1	1. Rxe6 Kg3 2. Rf6 Kh3 3. Rf3# 0-1	tg2500+photo
78	1b5b/8/Q7/7p/8/4p3/K7/2k1b3 w - - 0 1	1. Qd3 h4 2. Kb3 h3 3. Qc2# 0-1	tg1500+photo
79	8/3P4/8/8/8/k7/B2B1pB1/1K6 w - - 0 1	1. d8=Q f1=Q+ 2. Bxf1 Ka4 3. Qa5# 0-1	tg2500+photo
80	5n2/7P/8/B7/2N5/8/8/1k3K2 w - - 0 1	1. h8=Q Kc1 2. Qc3+ Kd1 3. Qd2# 0-1	comp3.5
81	2K5/k3N3/8/1B6/8/8/p7/8 w - - 0 1	1. Kc7 a1=Q 2. Nc8+ Ka8 3. Bc6# 0-1	comp3.5
82	4K3/8/8/8/Q3B3/8/7k/8 w - - 0 1	1. Qa3 Kg1 2. Qg3+ Kf1 3. Bd3# 0-1	tg2500+photo
83	7k/8/n4K2/8/8/r1p5/8/2R5 w - - 0 1	1. Kf7 c2 2. Rh1+ Rh3 3. Rxh3# 0-1	comp3.5
84	3k1K2/8/4P3/1B6/1R6/3r4/8/1b3R2 w - - 0 1	1. Rc4 Rf3+ 2. Rxf3 Bc2 3. e7# 0-1	comp3.5
85	5R2/6N1/8/5K2/7k/1b6/6p1/8 w - - 0 1	1. Kf4 Bf7 2. Rh8+ Bh5 3. Rxh5# 0-1	tg1500+photo
86	7N/P7/1B1k4/7K/3N4/8/8/8 w - - 0 1	1. a8=Q Ke5 2. Ng6+ Kf6 3. Qf8# 0-1	comp3.5
87	4R3/K2R1p2/8/8/8/8/8/6Nk w - - 0 1	1. Rg8 Kh2 2. Rxf7 Kh1 3. Rh7# 0-1	tg2500+photo
88	4B3/5Pk1/8/7P/7B/8/5b2/3K4 w - - 0 1	1. Be7 Bg3 2. f8=Q+ Kh7 3. Bg6# 0-1	tg1500+photo
89	8/8/5Q2/8/k3p3/8/K7/2N5 w - - 0 1	1. Qb6 e3 2. Nb3 e2 3. Nc5# 0-1	tg1500+photo
90	8/P7/8/8/2k5/2b5/2R5/1K1Q4 w - - 0 1	1. a8=Q Kb4 2. Qb7+ Kc4 3. Qdd5# 0-1	tg2500+photo

Appendix D
The Human Expert Evaluations of the 90 DSNS-Generated Compositions

#	Jana Krivec	Vlaicu Crisan	Matej Guid	Average
1	0.0	0.5	0.5	0.33
2	1.0	0.1	1.0	0.70
3	1.0	0.0	0.0	0.33
4	0.0	0.1	0.1	0.07
5	2.0	0.1	4.0	2.03
6	2.0	0.3	0.1	0.80
7	2.0	0.0	3.0	1.67
8	1.0	0.2	0.1	0.43
9	1.0	0.1	0.2	0.43
10	2.0	0.2	0.2	0.80
11	2.0	0.5	3.0	1.83
12	1.0	0.2	0.0	0.40
13	1.0	0.3	3.0	1.43
14	1.0	0.1	1.0	0.70
15	1.0	0.0	0.0	0.33
16	0.0	0.2	0.1	0.10
17	1.0	0.1	0.2	0.43
18	1.0	0.1	0.2	0.43
19	1.0	0.1	0.2	0.43
20	2.0	0.3	0.2	0.83
21	1.0	0.0	0.0	0.33
22	2.0	0.0	0.5	0.83
23	0.0	0.2	0.1	0.10
24	1.0	1.0	3.0	1.67
25	2.0	0.2	2.0	1.40
26	2.0	0.1	0.3	0.80
27	0.0	0.0	0.0	0.00
28	0.0	0.1	0.2	0.10
29	2.0	0.2	1.5	1.23

(continued)

A. Iqbal et al., *The Digital Synaptic Neural Substrate*,
SpringerBriefs in Cognitive Computation, DOI 10.1007/978-3-319-28079-0

#	Jana Krivec	Vlaicu Crisan	Matej Guid	Average
30	0.0	0.2	0.0	0.07
31	0.0	0.3	0.0	0.10
32	2.0	0.1	2.0	1.37
33	0.0	0.0	0.0	0.00
34	2.0	2.0	0.2	1.40
35	2.0	0.1	0.2	0.77
36	2.0	0.1	3.0	1.70
37	1.0	0.1	0.2	0.43
38	0.0	0.1	0.0	0.03
39	2.0	0.0	0.0	0.67
40	1.0	0.2	0.3	0.50
41	0.0	0.1	0.0	0.03
42	2.0	0.0	0.2	0.73
43	1.0	0.0	0.2	0.40
44	1.0	0.2	0.0	0.40
45	2.0	0.0	0.2	0.73
46	1.0	0.2	0.4	0.53
47	3.0	0.0	0.8	1.27
48	0.0	0.1	0.2	0.10
49	0.0	0.1	0.3	0.13
50	1.0	0.0	0.0	0.33
51	1.0	0.3	0.3	0.53
52	2.0	0.3	1.0	1.10
53	3.0	0.0	0.1	1.03
54	0.0	0.1	0.6	0.23
55	0.0	0.2	3.0	1.07
56	0.0	0.2	0.1	0.10
57	1.0	0.1	0.1	0.40
58	1.0	0.1	0.1	0.40
59	2.0	2.0	0.1	1.37
60	1.0	0.3	0.3	0.53
61	1.0	0.3	0.5	0.60
62	2.0	0.4	3.0	1.80
63	2.0	0.4	2.0	1.47
64	2.0	0.2	0.1	0.77
65	2.0	0.1	0.3	0.80
66	1.0	1.0	0.1	0.70
67	1.0	0.0	0.6	0.53
68	1.0	1.0	1.0	1.00
69	3.0	0.3	0.1	1.13
70	3.0	0.3	2.0	1.77

(continued)

#	Jana Krivec	Vlaicu Crisan	Matej Guid	Average
71	2.0	0.0	3.0	1.67
72	2.0	0.3	1.5	1.27
73	3.0	0.2	0.2	1.13
74	3.0	0.3	0.2	1.17
75	3.0	0.1	3.0	2.03
76	2.0	0.2	0.2	0.80
77	3.0	0.2	0.0	1.07
78	3.0	0.2	0.5	1.23
79	2.0	0.0	0.1	0.70
80	2.0	0.1	0.1	0.73
81	4.0	0.3	2.0	2.10
82	3.0	0.0	0.3	1.10
83	1.0	0.3	0.0	0.43
84	2.0	0.3	0.2	0.83
85	2.0	0.3	0.1	0.80
86	2.0	0.2	0.1	0.77
87	2.0	0.0	0.1	0.70
88	1.0	0.3	0.5	0.60
89	1.0	0.1	0.3	0.47
90	2.0	0.0	0.1	0.70

Appendix E
Unedited Expert Commentary on the DSNS-Generated Problems

Note that these comments were provided with no knowledge about the actual composer of the problems. The experts would have had to guess whether they were human or computer-generated. The parts in bold refer to the additional commentary the expert provided for the compositions he identified as most likely having been composed by a human. In the move notation 'S' and 'N' both refer to the knight.

#	FM (C) / IM (S) Vlaicu Crisan	FM Matej Guid
1	Very weak key takes three flights. Short threat. Two variations ending with grab theme and same mate as in the threat. No duals in the real play a plus	Obvious, plain
2	Awful key: major promotion from en prise position capturing a black officer with short threat and taking all four flights. wRc2 used only as threat. Dual in the variation 3.Qa4#	Slightly pretty, but with obvious moves **There are many pieces and pawns on board (harder to design a problem automatically), perception of beauty is relatively humanlike (although this could be very subjective)**
3	Double solution: 1.Qa7! ~ 2. h8=Q/R[+] ~ 3.Q[R]xf8#. The key takes three flights to the black King. The threat is unstoppable	Plain
4	Weak give-and-take key: en prise rook captures the black Queen (last officer). The dualistic mate 3.Qb2# ruins the intended beauty of a mirror ideal mate	Most straightforward

(continued)

A. Iqbal et al., *The Digital Synaptic Neural Substrate*,
SpringerBriefs in Cognitive Computation, DOI 10.1007/978-3-319-28079-0

#	FM (C) / IM (S) Vlaicu Crisan	FM Matej Guid
5	Short dualistic threat after the key. Three useless pieces (wRd4, wBc5 and bBbg7) - they could be easily replaced by white pawns d4 and e3. Poor construction	Fairly difficult to find (there are reasonable alternatives at disposal) … also beautiful in the sense that a quite march forward by the pawn results in an effective check(mate) on the long diagonal **This one was fairly complex, with a nice motif.. I also gave it the highest aesthetic rating… most computer-like compositions are far more simple (also in a computational sense)**
6	Bad key, capturing a whole Rook (last black piece) and taking two flights. Poor construction: bR can be replaced by bP. A second variation can be added easily (Ke3, Re1 - Kh2, pe4) **The wR could actually stand anywhere from a4 to a8. I think a4 would be the preferred choice of a human, because of the figurative initial setting**	Very obvious moves, hard to miss any one of them
7	Double solution: 1.Qg3! Kd4 2. Rh5 Ke4 3.Rh4#. Dual in the main variation 1.Qd2 Ke5 2.Rh3 Ke4 3. Qe3#. No dual in the second variation 1.Qd2 Kf3 2.Rh3+ Kg4 (Ke4) 3.Be6(Qe3)#	Nice geometric position of the pieces a the end… but the fact that the black king is the only black piece makes it easier to find the solution… still: pretty **It is possible that this one was designed by a computer… I find the geometry really pretty, and for this reason it seems to me that it is more likely that a human was the composer**
8	Weak key, taking two flights of which one is provided in the set play (e8) and short threat. No real fight: black must sacrifice its Bishop to stop the immediate mate. No duals	The strongest - and very obvious - continuation wins, which makes it less beautiful
9	Obvious key, capturing the remaining black piece and taking five flights. Duals after 1…Kf1 2. Qd2/Qg6/Qc2/Qf3. Usage of two white Queens rather dubious	Not capturing the knight would be beautiful, the solution involving capturing it is not

(continued)

#	FM (C) / IM (S) Vlaicu Crisan	FM Matej Guid
10	Key takes two flights and threatens two short mates. One variation only, with a line closing and line opening, ending with capture. Fortunately, no duals	Possible interposition of the black knight makes the solution less beautiful
11	Ampliative though pretty obvious key, giving an extra flight. However, all three black moves have the same continuation (2.Rb2 and 3.Ra1#) - so no real actual fight	A strange and somewhat unexpected white's knight move makes the solution beautiful, in particular since it is the only way to deliver checkmate in time **This 1.Nd1 is very appealing and unexpected… due to the small number of pieces it is quite possible that the computer designed it… but would computer know to appreciate the beauty of this move?**
12	Promotion key takes two flights. Long threat is actually unstoppable. bS defense only stops a second similar threat (2.Qc7+ Ka8/Ka6 3. Sb6/Qb6#)	A very obvious solution
13	Obvious key takes two prominent flights. Dual after 1…Kh2 2.Qf2/Qf3. Compare with Shinkman's composition (Ke4, Qb5, Ba5 - Kc8): 1.Qb2 Kd7 2.Qe5 Kc8(Kc6) 3.Qc7(Qd5)#	Nice geometry… the fact that this is the only way to deliver checkmate in time makes it quite pretty
14	Key takes bK flight, creates a strong unstoppable battery and threatens a dualistic mate (2.Sc6/Sd7) which Black defense can not actually prevent. wSe5 useless. No real fight	Sacrificing of the knight in order to enable a checkmate to the "determined" rook is not that plain
15	Three double solutions: 1.Kf5/Kh5/Kf6 all threatening 2.Rg8#. 1…Bc5 or 1…Kf8 can both be answered by 2.Rg7 followed by 3.Qf7#. Brutal key capturing the last black piece	An obvious route to checkmate
16	Dualistic mate 3.Qa4/Qa3/Qb2# are also possible. No real challenge for a player: the mate is possible only by approaching wK. Black has no defense against the threat	Plain march by the white knight… not beautiful

(continued)

#	FM (C) / IM (S) Vlaicu Crisan	FM Matej Guid
17	Duals in the main variation: 2.Bc1/ Bd2/Be3+. Extra variation is better: 1…Kf5 2.Be3 Ke5 3.Rg5#. Bad key capturing last black officer, taking a flight and setting up a strong battery	The other possible solution would be quite pretty: 1… Kf5 2. Be3 Ke5 3. Rg5# (but the actual continuation is not)
18	Dual in the main variation: 2.Rxb2 also works, followed by 3.Ra1#. The key takes a flight and threatens a short mate, which black can only delay. No duals with wKe7, bSc5 and wBa1	Not taking the en-prise knight is ok, but still very plain
19	Major promotion key takes two flights. Dualistic threat 2.Qf7+ also works. A secondary variation adds interest: 1…Kxf6 2.a8=Q Ke7 3. Qad8#	Nice to see the new queen promoted… but the rook move would deliver the checkmate as well… not really beautiful
20	Black pieces are rather superfluous - even without them, the same solution would appear (1.Kb6?? Stalemate). Solution is however obvious, because no real fight **A computer might have seen that bSe3 and bPe6 are useless. A human would probably add them just in order to equilibrate the balance of initial forces**	Too obvious
21	Double solution: 1.Be5! Sg3 2. Rxg3 Kh4/Kh6 3.h8=Q[R]#. Bad key taking a flight and threatening a short mate	Obvious, besides 1.Be5 is another solution
22	Double solution: 1.Rc7! Kh8 2.Rd8 Kg8 3.Se6/Sg6/Sh7#. Dual in the intention after 1…Kh8 2.Se6/Sg6+/ Rc8. Bad key taking two flights	Leaving the knight en prise is somewhat beautiful
23	Obvious key takes three flights. It would have been better to place bP on e4 in order to set up a continuous zugzwang: 1.Qg1! Zz 1…e3 2.Be2! Zz 2…Kh4 3.Qg4# **Human are perhaps more likely to create zugzwang based compositions**	Too obvious

(continued)

#	FM (C) / IM (S) Vlaicu Crisan	FM Matej Guid
24	Key takes flight c4 and threatens the variation given in the solution 2. Sd3+ Kb3/Ka3 3.Qb2#. Also particularly interesting is the second variation: 1…Bd4 2.Qxd4+ Ka3 3.Qc3# **Neat changed mate after 2…Ka3 in two variations is particularly pleasing**	Nice geometry… also the only way to deliver checkmate in time… quite pretty **The geometry is really not trivial… humans know to appreciate it… the quality seems to be much higher that in many of these 90 problems**
25	Key is bad, capturing a black piece that can give check and introduces a short threat. In the main variation, black second move can stop only one of the threatened mates. No duals	Nice geometric moves by the newly promoting queen… long moves
26	Dual after 1…Rd1+ 2.Sxd1. Flight taking key. wSe3 could be removed, by placing for instance wK on g2 and wRc4 on c5	Very obvious… not taking the rook is ok.. but still
27	Double solution: 1.Rxb2! ~ 2.Ra7/Rd7 ~ 3.Ra1/Rd1#. Dual in the intention after 2…c5 3.Rd1#. The threat is unstoppable - no real fight by Black	Not unexpected at all
28	Duals after 1…Sd3+ 2.Ke3/Kg3/Kg5. Key takes three flights and threatens a short mate. Poorly constructed position, with duals in the second variation: 1…Sg2+ 2. Ke5/Kg5/Kg3/Ke4	Ignoring the en-prise knight… still, too straightforward
29	Flight taking key, threatening a short mate. Black defense pins wQ, but the pin is released through a brutal capture of pinner. Duals: 2…a3 3.Qb7/Qb8#. Nothing subtle	Actually the only way by far… that's quite nice, but still not very beautiful **Lots of pieces and pawns makes it more difficult for the computers to compose**
30	Another horrible key, taking the most prominent black figure (bQ). 2…Bc6 ensures that 3.Qb7# is avoided. wSs can be traded for wPe2, when bQ is replaced on d3: 1.exd3!	Some checkmate combination with an underpromotion to a knight would be beautiful, capturing the black queen on the first move isn't
31	Straightforward play, with no surprises. There is a dual 2.b8=R which occurs after 1…g2. No need to have black pawns on the board **Black pawns d4 and g3 are seemingly useless**	Very plain

(continued)

#	FM (C) / IM (S) Vlaicu Crisan	FM Matej Guid
32	Multiple duals in intention after 1… Se5 2.Bb3/Bd5/Be6/Ba3/Bb4/Bc5/Kc8. The flight taking key is not pleasant	The helpless black knight, a unique solution
33	Double solution: 1.Rc6! with the following variations: 1…Rh8/Rh6 2.Re7+ Kf8 3.d8=Q# and 1…Kf8/Kg8/Kf7 2.d8=Q[+] ~ 3.Re7#. Bad key taking three flights	The most obvious solution
34	The second variation 1…Kf8 2.Sg6 + Kg8 3.Rg7# enhances the content of this composition. The key is not bad and a better one can not be found. Best from the lot **This is clearly the best of the lot. A human composer would even consider publishing this composition!**	Too simple
35	Duals: 2.bxa8=R/B/S/Kxa8. The second variation 1…Re2 2.h7 adds some interest. The key takes an unprovided flight and creates a short threat	1… Re2 2. h7 with an overloaded black rook would be slightly more pretty… just giving the rook away is not
36	Duals: 3.Qf5# and 2.Kd3 Kf3(Ke5) 3.Rf6(Qe4/Qg5)#. With same material, Healey created a memorable composition (Kd6, Qf2, Rc5, pc3 - Kd3): 1.Kd7 Ke4 2.Rd5 Kxd5 3.Qd4#	The pawn unexpectedly joins in the checkmating construction.. also the only way for #3… quite nice **The role of the pawn in this checkmate in terms of making this composition appealing has to be appreciated… seems humanlike perception of beauty**
37	Duals in the main variation: 3.Qb6/Qc8# and 2.d8=R ~ 3.Rdc8#. The key is not very good, because wR is en prise in the initial position	Too obvious
38	Dual: 2.Bd8 (instead of 2.Bf4+), avoidable by shifting wBc7 to b8. Again a major promotion key taking three flights. Second variation 1…Kh4 2.Bf4/Bd8 Kh5 3.Qg5#	Too obvious
39	Double solutions: 1.Rh1! Bb1 2. Rxb1 Ka5/Ka3 3.Ra1# and 1.Rh8! Be8 2.Rxe8 Ka5/Ka3 3.Ra8#. Duals in the intention after 1…Ka5 any waiting move e.g. 2.Kf7/Rb8/Rc6 etc. will mate	Too obvious

(continued)

#	FM (C) / IM (S) Vlaicu Crisan	FM Matej Guid
40	Dual after 2…Kh4 3.Rh7#. Give and take key introducing a short threat, with wR playing from en prise position	1. Rg8 is not the first move that comes too mind.. that slightly contributes to the aesthetic value
41	Multiple duals: 2.Rh6/Rf6/Sc3/Se3. The key is awful, capturing the only black unit left on the board (bQ). Everything is very obvious	Too straightforward
42	Double solution: 1.c8=R! Ka3 2. Rc4 Ka2 3.Ra4#. Nice stalemate avoidance in the intention 2.Qc4?? **In spite of the double solution, the stalemate 2.Qc4 would appeal much to humans, as it reminds the famous Barbieri-Saavedra endgame**	Underpromotion to a rook would be more pretty… 2. Qg4 is not too unexpected
43	Double solution: 1.Rd6! Kf1 2. Rd2/Bf3 Ke1 3.Rd1#. Dual in the intention as well: 2.Rg6 Ke1 3. Rg1#. Bad key taking an unprovided flight	Too obvious
44	Capturing key, creating a short threat. The defense is unique, but the continuation is rather dull. With wR placed on g4 and bPs a6, b7 and h4 removed, the key would have been improved **Black pawns a6 and b7 are seemingly useless. As stated in the commentary, the position could have been improved**	An obvious choice
45	Double solution: 1.Kh6! e3 2.Kg6/ Ba2-f7[Bh2] e2 3.Bd4[Be5]#. Similar duals in the intention: 2. Kh6/Ba2-f7. Bad key taking an unprovided flight	Most obvious
46	Dual: 3.f8=R#. Flight taking key. The play could have been improved by adding a twin (e.g. Shift bK to h8 with the solution: 1.Be4 Kg8 2. Ke7 Kh8 3.f7#)	There is only one possible #3 solution… and that is slightly surprising.. but fairly easy one to find
47	Double solutions: 1.Kb7/Kc7! Kb5 2.Bc2 Kb4 3.Qa4#. Duals in the intention: 2.Kb7/Kc7. Also dual after 2…Kb4 3.Qc5#. Key piece out of play in the initial position	Nice to engage the queen in such geometry… there are however several alternative solutions

(continued)

#	FM (C) / IM (S) Vlaicu Crisan	FM Matej Guid
48	After a major promotion key taking a provided flight, the play is dualistic: 1…Sf8 2.Qh5+/Qe1+/Qe4+/Qe7+/Rh2+/Qe3/Qe5/Qf7/Qb8/Qc8/Qxf8/. Same apply to other variations	So many alternatives on the 2nd move
49	Major promotion key takes three flights. Duals in the main variation after 2…Ke5 3.Ra5# or 1…g3 2.Ra6+/Kd4/Ke4. Spoilt by duals in all variations	Nice geometry… but certainly not the unique solution
50	Double solutions: 1.h8=R! ~ 2.Rg8+ Kh1 3.Rf1# and 1.Kg3! ~ 2.h8=Q ~ 3.Qh2#. Duals in the intention: 2.Qg7+/Qxe5/Kg3. Bad key (major promotion) and several threats	Most obvious
51	Promotion key takes a provided flight and introduces a short threat. Just one line, with no duals	Unique, but too obvious solution
52	Give (e5) and take (c6, e6) key by an en prise piece. The rest is forced, in spite of what black moves. No duals	The mating net by the knight and bishop is nice
53	Interversion of moves possible: 1.Rb4! Kc5 2.Rgb6 Kd5 3.Ra5#. Double solution: 1.Ra4! Kc5 2.K any (or R waiting) Kd5 3.Rb5#. Dual in intention: 2.Ra4 Kd5 3.R2b5#	Too elementary
54	Duals in the main variation: 1…Bb6 2.Qd5+/Qg5+/Qd7+ and 2…Kb4 3.Ra4/Rc4#. Again major promotion key takes flight and threatens a short mate	Diagonal checks on white squares vs. the opposite color bishop is a nice choice, but still fairly plain
55	Dual in the main variation: 3.Qb5#. The key played by en prise wQ threatens a short mate (2.Sc4#). The alternate black defense 1…Se3 is dualistic: 2.Qc7+/Bd3+/Bd7+/Bf1+/Be2+/Bc4+	Quite pretty **Relatively complex… although I wouldn't be too surprised if the computer composed this one (the absence of pawns reduce the complexity in the computational sense)**
56	Key takes black piece with major promotion. Dual 2.Qb1 after 1…Kd2. Duals also after 1…Ke3 2.Qb2/Qd8 - not only 2.Qb3+. Not exciting play	Too obvious

(continued)

#	FM (C) / IM (S) Vlaicu Crisan	FM Matej Guid
57	Flight taking key, capturing the only black officer and threatening two short mates. Dual after 1…Kh4 2.Qg5+/Qxg6/Sf2	Too obvious… e.g. a unique solution without capturing the knight (at least not on the first move) would be much prettier
58	Dual 2.Kc3 in the main variation. Bad key, taking three flights. Again this must be compared with Shinkman's cited at no 13	Obvious
59	Key captures the remaining black officer, but provides a flight. Two ideal mirror echo mates delivered thanks to zugzwang, the second after 1…Kc6 2.Rb4 Kd6 3.Rb6# **Two echo variations - something that humans will appreciate and try to create**	Plain
60	Duals: 3.Rh5/Rh6#. Other not dualistic variations: 1…Kg6 2. Rxc6+ Kh7 3.Re7/Rh5/Rh6# and 1…Rxb6 2.Qf7+ Kh6(Kh8) 3.Rh5 (Re8)#. Key takes three flights, though.	Obvious… still, at least the rook wasn't captured immediately
61	Key takes three flights, defending the wS en prise. No real fight: white threat can not be actually stopped	Obvious, but looks efficient
62	Only two flights taken by key - defect would have been avoided by putting wQ initially on e7 (or Kb4 Qf6 Sf3 - Kd5; B: wQf6 -> e8). Continued zugzwang and midboard mate **An appealing diagonal mate. An equally appealing setting would be Kc4 Qd4 Sh4 - Ke4**	Nice geometry, unique solution… quite pretty **Very appealing… the quality of this problem seems to be much higher than of the average problem here… hence the decision**
63	Single line, but key is obvious, defending wR en prise. Black active distant selfblock is exploited in the mate **For the nice symmetry - asymmetry**	Nice geometry, unique solution, minor pieces… pretty.. still relatively obvious (hard to decide whether or not it was composed by a human) **It was quite difficult to decide on this one… still, the minor pieces combine an appealing net… and it is probably relatively difficult computationally to operate with this subset of pieces**
64	Flight taking first move followed by flight taking second move with imparable mate. Pretty ordinary combination, often seen in practice	Obvious

(continued)

#	FM (C) / IM (S) Vlaicu Crisan	FM Matej Guid
65	Major promotion threatens three short mates. Main variation spoilt by many duals: 1…Bce7 2.Qb3+/Qb5+/Qb7+/Qb6/Qc7/Qc8. Second bB not justified	Diagonal check on white squares vs. the opposite color bishops.. but still not really beautiful
66	Clear single line, with forced continuation. Key takes flight and stops pawn's advance. bPe5 used in order to avoid stalemate **Just for the fact that Re4-c4 and Rc4-e4 are equipollent vectors and Kb1, Bc2, Re4 are aligned on the same diagonal. Many similar settings are possible**	Very straightforward
67	Double solutions: 1.Qg1/Qg6! with two unstoppable threats: 2.Qg2+ Kh4 3.Qg4# and 2.Qg4+ Kh2 3.Qg2# (also 1.Qf2 cooks). Duals in the intention: 2.Qc7/Qf2/Qg1. Flight taking key	Mating net with bishop and knight.. the second move (Qb8) is certainly more beautiful than 2.Qf2
68	Nice minor promotion, but bad key taking flight. A better presentation of this idea was shown by Mackenzie (Kc6, Bb8, pc7 - Ka8): 1.Ba7 Kxa7 2.c8=R Ka6 3.Ra8# with stalemate avoidance **For the minor promotions. Human even composed the following: Ke4 pc7 d6 e7 f6 g7 - Ke6: 1.e8=B Kxd6(Kxf6) 2.c8=R (g8=R) Ke6 3.Rc6(Rg6)#. Are computers able to do something similar?**	Underpromotion to a rook is nice… also this is a unique solution… with so little pieces.. quite pretty.. although easy to find
69	Dual 3.Qa6#. Again the key takes an unprovided flight and the second move captures the remaining officer, hence putting black in zugzwang	Not capturing the knight on the first move… still, an obvious solution
70	No surprise: key takes flight, second move takes main black officer and unavoidable mate follows. No real fight	A unique solution… each piece takes part.. black succeeds promoting to a queen, still helpless against the timely mate… pretty (and quite humanlike if it was composed by the computer) **Such geometry is something that humans know to appreciate (and compose)**

(continued)

#	FM (C) / IM (S) Vlaicu Crisan	FM Matej Guid
71	Double solution: 1.d8=S! ~ 2.Ba7/Bb7/Bf2/Bg1 ~ 3.Sc7/Sf4#. In the intention Black can not parry the threat - actually no real fight	Many alternatives for black, but Qd8-c8-c4 is always decisive in time… quite pretty! **The fact that the same maneuver wins in all variations makes me believe that it is very likely that a human composed this one**
72	En prise officer takes the remaining black piece, which was almost promoting and threatening to capture another white officer. The rest is forced, with interesting pawn mate	Pretty… it's a pity that the pawn must be taken!
73	Typical ending for normal game: flight taking key, followed by a flight taking second move with unavoidable mate. Lone bK hunting	Obvious
74	Key takes three flights. Black is in zugzwang and white plays 2.Rg2 and 3.Qe2# regardless black's moves. Nothing really exciting	Not so trivial… but also not very beautiful
75	Duals 2.Qd3+/Qe4+/Qf5. Flight taking key guarding also key squares a6 and d3	This one is nice… not easy, unique solution, geometry **With these pieces there are many possible checkmates, but this one is quite appealing… and that makes it more likely in my opinion that a human composed it**
76	Key takes bR threatening wQ. Neat mate in the main variation, similar with 1…Kc4. Dual after 1…Kb4 2.Kb2/Kc2 Kc4 3.Ra4#. Black Rook could be replaced by black S/B **For the appealing diagonal mate (like 62). Another appealing setting would be Kc1 Qc2 Bc5 - Kc3**	Obvious
77	Key takes bR threatening wR and also three flights. Similar second variation: 1…Kf2 2.Rb3 Kf1 3.Rf3#. Black Rook could be removed when wRe6 is shifted initially on b6 **Mirror ideal echo mates - long time humans considered them a symbol of perfection. Nowadays, however, strategic school prevails**	Most straightforward

(continued)

#	FM (C) / IM (S) Vlaicu Crisan	FM Matej Guid
78	Horror: two promoted black Bishops can't do anything against the threat. Key takes three flights. No real fight	White diagonals vs. black colored bishops… quite pretty
79	Double solutions: 1.d8=R! f1=Q+ 2.Bxf1 Ka4 3.Ra8# and 1.Bf1! Ka4 2.d8=Q[R] Ka3 3.Q[R]a8#. Dual in the intention: 3.Qa8#. The key threatens two short mates. No real fight from Black	Promoting to queen is obvious
80	Duals in the main variation 3.Sb2/Se3# and 2.Qb2+/Sa3/Ke2. Major promotion threatens short mate and takes flight	Straightforward
81	Typical endgame mate. Key takes flight and threatened mate can't be avoided. With similar material better composition were shown (Ke7 Ba1 Sh6 - Kh8 pg7 h7 B: Rotate table 180) **A typical position to be shown by humans, especially when explaining the finale KBS versus K**	Just in time mate… unique solution (actually, also the only one that wins)… quite nice **So efficiently pretty… the knight must go to a square previously occupied by the king, which makes the solution a bit harder to find… that may have been one idea of the (human) composer**
82	Double solution: 1.Qb3! Dual in intention 2.Qf3 Kh2 3.Qg2#	Plenty of solutions, and this geometric one is certainly not the ugliest
83	Key takes flight g8 and threatens short mate. No black defense can change anything, except 1…Ra1 2. Rxa1 and 3.Rh1#. Mate on the file is often met	Obvious
84	Key takes two flights and threatens short mate. Other variations: 1… Re3 2.Rf7 and 1…Rd7 2.Bxd7/exd7	Not most obvious… not too pretty either
85	Key takes flight g3 and threatens short mate on the file. Other variation: 1…Bg8 2.Rxg8	Straightforward
86	Key takes flight thanks to major promotion. Short mates if black plays other moves. Dual after 2… Kd6 3.Qc6/Qd8#	Promoting to queen on the first move is generally not the prettiest way… also, white has quite a lot of pieces
87	Double solutions: 1.Rxf7! ~ 2.Rg8 ~ 3.Rh7#; 1.Rd2! ~ 2.Re1 ~ 3. Sf3/Sh3# and 1.Re2! ~ 2.Rd1 ~ 3.Sf3/Sh3#. Duals in intention: 2. Rd3/Rd4 ~ 3.Rh3/Rh4#. Bad key taking a flight	Not capturing the pawn would be somewhat prettier… the solution is very obvious

(continued)

#	FM (C) / IM (S) Vlaicu Crisan	FM Matej Guid
88	Key takes flight, with wB leaving from en prise position. The long threat is actually unavoidable. However the line opening for wBe8 is neat	The moves are obvious, but the setup is relatively pretty **Perhaps this one is not so convincing… could have been a computer, the solution is somewhat too obvious… not a composition of a high quality.. still, interesting for beginners of chess**
89	Duals in the main variation: 3.Qa5# and 2.Sd3 also work. Key immobilizes bK, by taking three flights	Efficient, but obvious
90	Double solution: 1.a8=R! [2.Rb8 Kc5 3.Rxc3#] 1…Kb5/Kc5 2.Rxc3 [+] ~ 3.Qb3# 1…Kb3 2.Rb2+ Kc4 3.Rc8# and 1…Kb4 2.Qd5 B~ 3. Qb7/Qc4#. Duals in intention: 2. Qa6/Qd5/Qd6+	Lots of alternatives, very obvious

Appendix F
Unedited Microsoft Visual Basic 6 Source Code of Chesthetica v9.95's Composing Subroutine

Note that the code in this subroutine, though readable, is umedited and inelegant as it was improved upon and enhanced in its functionality over several years. It nevertheless works and illustrates the thought processes and myriad other issues that factor into an actual composing program. This code also incorporates the composing of four-movers, five-movers and studies. Comments are preceded by an apostrophe.

A. Iqbal et al., *The Digital Synaptic Neural Substrate*,
SpringerBriefs in Cognitive Computation, DOI 10.1007/978-3-319-28079-0

```
Public Sub Compose_Mate()
'composes mate-in-3, mate-in-4, mate-in-5 and study compositions automatically

If code_viewer_loaded = True Then Call Code_View.Update_Code("Public Sub Compose_Mate")

Dim a1 As Integer, b1 As Integer, c As Integer, d As Integer, e As Integer, i As Integer, j As Integer, k As Integer,_
l As Integer, m As Integer, n As Integer, o As Integer, repeat_count As Integer
Dim piece_chosen As Integer, prnm As Integer, remaining_prob As Integer, wb_adjust As Integer, _
temp_piece As Integer, chosen_technique As Integer, pieces_w As Integer, pieces_b As Integer, kpm_initial As
Integer, _
kpm_after_m1 As Integer, piece_counter_val As Integer, square_counter As Integer, pcount_failsafe As Integer
Dim temp_board() As Integer, sqrs_arnd() As Integer
Dim Mi3s_in_PGN As Double, composing_efficiency As Double, blank_prob As Double, total_aes As Double, _
avg_aes As Double, h_aes As Double
Dim white_piece As Boolean, castling_flag As Boolean
Dim strtext As String, file_name As String, temp_gcm As String, piece_color As String, optimized_GCM As String
Dim arlines() As String, temp_fen As String
Dim time_to_start As Variant
Static total_attempts As Double, prev_total_attempts As Double, comp_success As Double
Static exp_tst As Variant  '(experiment) time to start
Static main_cap_length As Integer
Dim dsns_constraints As Constraints
Dim starting_time As Variant, starting_time_original As Variant, ending_time As Variant
Dim ij As Integer, eco_rounds As Integer
Dim composed_this_cycle As Double, composition_efficiency As Double, composed_thus_far As Double
Dim aesthetics_before_optimization As Double, aesthetics_after_optimization As Double
Dim current_fen As String, ftp_server_bckup As String
Dim is_ce As Double  'intelligent selection composing efficiency
Dim elapsed_time As Double, factor As Integer, most_aesthetic_lm_line As String
Dim LML As Integer  'long mate length
Dim long_mate_aesthetics As Double
Static long_mate_compositions As Integer
Dim wpermute_satisfied() As Integer, bpermute_satisfied() As Integer
Dim wp_counter As Integer, bp_counter As Integer, wp_choice As Integer, bp_choice As Integer
Dim v, w As Integer
Dim sparsity_value As Double
Dim pta As Integer 'white or black piece to add
Dim ptc As Integer 'piece type counter
Dim placement As Integer
Dim sum_year As Double
Dim ptc_limit As Integer
Dim gcm_strings() As String
Dim gcm_counter As Integer
Dim gcm_match As Boolean
Dim gcm_i As Integer
Dim mbp As Double  'material balance probability
Dim x As Integer
Dim swp() As Integer   'squares with pieces
Dim swpc As Integer    'swp counter
Dim sel_pl As Integer  'selected piece location
Dim ptr As Integer     'piece to remove
Dim original_gcm As String
Dim discounted_aesthetics As Double
Dim suspend_logical_filter As Boolean
Dim arlines_study() As String, temp_study_line As String, cp_value As Integer, algebraic_line As String
Dim forsythe_backup As String, study_aesthetics As Double
Static study_compositions As Integer


factor = 1 'this is used with DSNS + IS mode

If mnuOptAR.Checked = True Then mnuOptAR.Checked = False
'the animate replay function should be disabled during the composing process

If DSNS_Experiment = False And ER_Experiment = False Then
```

```
'these details are recorded only if an experiment is not running; for if it is, another file called 'Experimental
Results' is produced
'this part of the code is only visited once during the composing process so the existence of the file does not need
to be done each time
   Open app.Path & "\" & "Composition Specifications.txt" For Output As #48
   Call Register_File_for_Delete_Menu("Composition Specifications.txt", True)
   Print #48, "Started: " & Date & vbTab & Time
   Print #48,
   If Composer.Option1 = True Then Print #48, "Composing Technique: Experience"
   If Composer.Option2 = True Then Print #48, "Composing Technique: Random"
   If Composer.Option3 = True Then
      If Composer.ISL.value = 1 Then
         Print #48, "Composing Technique: DSNS + IS"
      Else
         Print #48, "Composing Technique: DSNS"
      End If
      Print #48, DSNS_source_files
      Print #48, Composer.dlgOpenFile3.FileName
      Print #48, Composer.dlgOpenFile4.FileName
   End If
   Print #48,
   If Composer.No_Cooks.value = 1 Then Print #48, "No Cooked Problems"
   If Composer.No_Duals.value = 1 Then Print #48, "No Duals"
   If Composer.No_Checks.value = 1 Then Print #48, "No Checks in Key Move"
   If Composer.No_Captures.value = 1 Then Print #48, "No Captures in Key Move"
   If Composer.No_Restrict.value = 1 Then Print #48, "No Restricting Enemy King Movement in Key Move"
   If Composer.No_Promo.value = 1 Then Print #48, "No Promotion to Queen"
   If Composer.MTOL.value = 1 Then Print #48, "Contain More than One Line"
   If Composer.DMEL.value = 1 Then Print #48, "Different Mate Each Line"
   If Composer.NSMP.value = 1 Then Print #48, "No Shorter Mates Possible"
   If Composer.CMAL.value = 1 Then Print #48, "Choose Most Aesthetic Line"
   If Composer.st.value = 1 Then Print #48, "Switch Techniques"
   If Main.mnuOptCompTM.Checked = True Then Print #48, "Compose Three-Movers"
   If Main.mnuOptCompLM.Checked = True Then Print #48, "Compose Longer Mates"
   If Main.mnuOptCompS.Checked = True Then Print #48, "Compose Studies"
   Print #48,
   Print #48, "Engine Solving Time: " & Composer.Solving_Time.Text & " s"
   Print #48, "Min Aesthetics Score: " & Composer.Min_Aesthetics.Text
   Print #48, "Logical Filter: " & Trim(Composer.Logical_Filter.Text)
   Print #48,
   If Composer.AO.value = 1 Then Print #48, "Piece Count: Optimize "
   If Composer.Cplx.value = 1 Then Print #48, "Piece Count: Complex"
   Close #48
End If

If Composer.CMAL.value = 1 Or Composer.ISL.value = 1 Then
   If Composer.Min_Aesthetics = vbNullString Then
      Composer.Min_Aesthetics = "0"
      'these two options rely on an aesthetics score being specified ('0' by default)
   End If
Else
End If

file_name = app.Path & "\" & "Generated Compositions.pgn"
If Filexists(file_name) = True Then
'determines how many compositions are already in the 'Generated Compositions' file (if there)
   Open file_name For Input As #29
   Mi3s_in_PGN = Word_Frequency(file_name, "FEN"): Close #29: Composer.Total_Mi3s = Mi3s_in_PGN
End If

If Filexists(app.Path & "\" & "Generated Compositions (Long Mates).pgn") = True Then
'determines how many compositions are already in the 'Generated Compositions (Long Mates)' file (if there)
   Open app.Path & "\" & "Generated Compositions (Long Mates).pgn" For Input As #59
   Composer.LMC_Old.Text = Val(Word_Frequency(app.Path & "\" & "Generated Compositions (Long
   Mates).pgn", "FEN")): Close #59
```

```
End If

    If Filexists(app.Path & "\" & "Generated Compositions (Studies).pgn") = True Then
     'determines how many compositions are already in the 'Generated Compositions (Studies)' file (if there)
       Open app.Path & "\" & "Generated Compositions (Studies).pgn" For Input As #59
         Composer.StudC_Old.Text = Val(Word_Frequency(app.Path & "\" & "Generated Compositions (Studies).pgn",
         "FEN")): Close #59
End If

   If Main.Caption = "Chesthetica " & About.lblVersion.Caption Then
         Main.Caption = Main.Caption & " - " & "*Composing Problems*": main_cap_length = Len(Main.Caption): Call
         Update_Tray_Tooltip
End If

   If Composer.Option1 = True Then chosen_technique = 1
   If Composer.Option2 = True Then chosen_technique = 2
   If Composer.Option3 = True Then chosen_technique = 3

  If Composer.Tmax.Text <> vbNullString Then
       'sets the upper limit of 'transformations' to a position to equal its maximum number specified or 30 (minus kings),
       by default 'this keeps transformations from continuing ad infinitum or the position freezing because it can no longer be

modified
       '50% allowance is provided to account for some illegal positions and piece shifting that may erase other pieces _
    (preventing the position reaching the maximum piece count specified)
       d = (Val(Composer.Tmax.Text) - 2) + ((Val(Composer.Tmax.Text) - 2) / 2)
   'adjusted for the kings
Else
   d = 30 + 15
End If

  starting_time = Date + Time

  If chosen_technique = 3 Then

      Call Extract_Excel_Data(Excel_DSNS_File_Name)
          'this extracts the Excel data one time so that it does need to be read from the 'physical' Excel file on the drive
         too often 'the global array, 'raw_excel_array' now contains all the relevant data from the Excel file


      If Excel_DSNS_File_Name2 <> vbNullString Then
         Call Extract_Excel_Data2(Excel_DSNS_File_Name2)
        'for the other domain Excel data
   End If
End If

1

   If Val(Composer.Composing_Time) <> 0 Then
        'this keeps the composer running for a fixed period of time until it stops automatically (e.g. in experiments
        running for 12 hours)
          If Val(DateDiff("s", Composer_Start_Time, Date + Time)) >= Val(Composer.Composing_Time) * 60 Then
        Call Composer.Stop_Button_Execute
   End If
End If

  Call Clear_the_Board
   final_mate = False  'resets the flag
     pcount_failsafe = 0 'resets the piece count failsafe flag

  If total_attempts >= 1 Then
        'the efficiency should only be calculated against *completed* attempts; not including the one in progress
        If Val((DateDiff("s", Composer_Start_Time, Date + Time))) <> 0 Then
             composing_efficiency = RoundIt((Val(Composer.Mi3s.Text) + long_mate_compositions +
             study_compositions) _ / ((Val((DateDiff("s", Composer_Start_Time, Date + Time))) -
             Total_Paused_Time) / 3600), 2)

   End If
```

```
        Composer.Comp_Eff = composing_efficiency & " cph"
        Composer.Comp_Eff.ToolTipText = "compositions per hour"
          Main.Caption = Mid(Main.Caption, 1, main_cap_length) & " - " & composing_efficiency & " cph"
           'this enables the progress percentage to be viewable as a tooltip when the program is minimized to the taskbar
           Call Update_Tray_Tooltip
End If

  If Composer.Rand_PC.value = 1 Then
        Composer.Wmin.Text = Generate_Random_Number(2, 4)
          Composer.Wmax.Text = Generate_Random_Number(1, 12 - Val(Composer.Wmin.Text)) +
          Val(Composer.Wmin.Text)
        Composer.Bmin.Text = Generate_Random_Number(2, 4)
          Composer.Bmax.Text = Generate_Random_Number(1, 12 - Val(Composer.Bmin.Text)) +
          Val(Composer.Bmin.Text)
         If Val(Composer.Bmin.Text) > Val(Composer.Wmin.Text) Then GoTo 1
          Composer.Tmin.Text = Val(Composer.Wmin.Text) + Val(Composer.Bmin.Text)
          Composer.Tmax.Text = Val(Composer.Wmax.Text) + Val(Composer.Bmax.Text)
          d = (Val(Composer.Tmax.Text) - 2) + ((Val(Composer.Tmax.Text) - 2) / 2)
End If

        total_attempts = total_attempts + 1: prev_total_attempts = prev_total_attempts + 1
        piece_chosen = 0: castling_flag = False
          Castling_Type(0).value = 0: Castling_Type(1).value = 0: Castling_Type(2).value = 0: Castling_Type(3).value = 0

 If chosen_technique = 3 Then
          'the FIRST part (pertains to the number of pieces of each color that can be used)
          'if the DSNS composing technique was chosen, the constraints need to be obtained

          dsns_constraints = Generate_Constraints_for_Composition(Excel_DSNS_File_Name,
          Excel_DSNS_File_Name2)
         'invoke the contraints generator using a particular set of compositions
          'this GSfC function knows to access the global Excel data array (raw_excel_array)

            Do 'prevents proceeding and hanging until the generation of constraints has been completed (see function for
the flag as 'false')
              'no doevents here because the MSUCI engine (and the creation of the analysis.txt file for checking forced
mate) sometimes _
        creates a conflict
            'if you allow doevents it will allows those things to invoke before this is done
        Loop Until dsns_constraints.completed = True

         Composer.Tmin = vbNullString: Composer.Tmax = vbNullString
      'removes the default values if any

         If Int(Val(dsns_constraints.number_of_white_pieces_s1)) <=
   Int(Val(dsns_constraints.number_of_white_pieces_s2)) Then
            'attribute minimum and maximum values for the white pieces based on the relevant attribute in the two newly
generated DSNS strings
            Composer.Wmin = Int(Val(dsns_constraints.number_of_white_pieces_s1))
            Composer.Wmax = Int(Val(dsns_constraints.number_of_white_pieces_s2))
  Else
            Composer.Wmax = Int(Val(dsns_constraints.number_of_white_pieces_s1))
            Composer.Wmin = Int(Val(dsns_constraints.number_of_white_pieces_s2))
  End If
         If Int(Val(dsns_constraints.number_of_black_pieces_s1)) <=
   Int(Val(dsns_constraints.number_of_black_pieces_s2)) Then
            'attribute minimum and maximum values for the black pieces based on the relevant attribute in the two newly
generated DSNS strings
            Composer.Bmin = Int(Val(dsns_constraints.number_of_black_pieces_s1))
            Composer.Bmax = Int(Val(dsns_constraints.number_of_black_pieces_s2))
  Else
            Composer.Bmax = Int(Val(dsns_constraints.number_of_black_pieces_s1))
            Composer.Bmin = Int(Val(dsns_constraints.number_of_black_pieces_s2))
  End If

          Composer.Tmin = Val(Composer.Wmin) + Val(Composer.Bmin) 'computes the total values accordingly
```

```
Composer.Tmax = Val(Composer.Wmax) + Val(Composer.Bmax)

'the SECOND part (generates the piece permutations that satisfy the white and black Shannon value
requirements)
Dim awsv As Integer, absv As Integer, p As Integer, q As Integer
'average white/black Shannon value
Dim wpermute() As Integer, bpermute() As Integer
'local arrays for the piece permutations (for each color)

awsv = Int(Round((Val(dsns_constraints.value_of_white_pieces_s1) +
Val(dsns_constraints.value_of_white_pieces_s2)) / 2))
absv = Int(Round((Val(dsns_constraints.value_of_black_pieces_s1) +
Val(dsns_constraints.value_of_black_pieces_s2)) / 2))
'these will be the total Shannon value upper and lower limits of the white and black pieces that can be used to
compose 'since there are two DSNS strings, each with white and black total Shannon values, the average is used


If new_dsns_strings_invoked = True Or Restarted_Composing = True Then
'this is necessary to allow the program to continue composing once the composing has been stopped and then
started again 'otherwise, the composing window has to be closed and the program restarted
  Restarted_Composing = False    'resets the flag
  Call Piece_Permutations(awsv)
  'calculates the piece combinations that satisfy the average white Shannon value
  ReDim wpermute(0 To UBound(PPermute), 0 To 4)
  'this block transfers the values from the global array to the local array (for the white pieces)
  For p = 0 To UBound(PPermute)
    For q = 0 To 4
      wpermute(p, q) = PPermute(p, q)
    Next q
  Next p

  dsns_num_of_tries = UBound(PPermute) + 1

  Erase PPermute  'erases the global array

  Call Piece_Permutations(absv)
  'calculates the piece combinations that satisfy the average black Shannon value
  ReDim bpermute(0 To UBound(PPermute), 0 To 4)
  'this block transfers the values from the global array to the local array (for the black pieces)
  For p = 0 To UBound(PPermute)
    For q = 0 To 4
      bpermute(p, q) = PPermute(p, q)
    Next q
  Next p

  dsns_num_of_tries = dsns_num_of_tries + UBound(PPermute) + 1
  'so the sum of tries is for both black and white permutations
End If

'so by this point, we have 1) the minimum and maximum number of white and black pieces that can be used,
and 2) the piece permutations _ that satisfy the average total Shannon value of those pieces (for white and black)


'the THIRD part (determines which piece permutations satisfy the first and second parts above, along with
having a reasonable _ number of each piece type to resemble a real composition)
wp_counter = 0: bp_counter = 0: wp_choice = 0: bp_choice = 0

ReDim wpermute_satisfied(0 To UBound(wpermute), 0 To 4)
ReDim bpermute_satisfied(0 To UBound(bpermute), 0 To 4)
'resizes the 'satisfied' arrays to match the w/bpermute ones, the counters will function as stop points later

For p = 0 To UBound(wpermute)
```

```
        If (wpermute(p, 0) + wpermute(p, 1) + wpermute(p, 2) + wpermute(p, 3) + wpermute(p, 4) >=
Val(Composer.Wmin) - 1) _
        And (wpermute(p, 0) <= 2 And wpermute(p, 1) <= 3 And wpermute(p, 2) <= 3 And wpermute(p, 3) <= 3) _
        And (wpermute(p, 0) + wpermute(p, 1) + wpermute(p, 2) + wpermute(p, 3) <= 8) Then
        'the total number of white pieces that can be chosen from should exceed the minimum piece requirement
(which includes the king)
        'the maximum piece requirement is irrelevant and will be controlled for automatically later in the code
        'also, only ONE or perhaps two of the pieces (other than pawns) can have an extra from the original set
(compositions tend to be _
        realistic)
            For q = 0 To 4
                wpermute_satisfied(wp_counter, q) = wpermute(p, q)
                'adds the permutation that satisfies the requirement
            Next q
            wp_counter = wp_counter + 1
        End If
    Next p

    If wp_counter > 0 Then
    'assuming at least one permutation that satisfies the requirement was found (quite likely)
        wp_counter = wp_counter - 1 'arrays start at 0
        wp_choice = Generate_Random_Number(0, wp_counter)
        'a random one is selected
         For q = 0 To 4
        'and the values filled in the Composer form
            Composer.WQS.Text = wpermute_satisfied(wp_choice, 0)
            Composer.WRS.Text = wpermute_satisfied(wp_choice, 1)
            Composer.WBS.Text = wpermute_satisfied(wp_choice, 2)
            Composer.WNS.Text = wpermute_satisfied(wp_choice, 3)
            Composer.WPS.Text = wpermute_satisfied(wp_choice, 4)
        Next q
    Else
        total_attempts = total_attempts - 1
        GoTo 1  'or the composing attempt has to start all over again
    End If

    For p = 0 To UBound(bpermute)
        If (bpermute(p, 0) + bpermute(p, 1) + bpermute(p, 2) + bpermute(p, 3) + bpermute(p, 4) >=
Val(Composer.Bmin) - 1) _
        And (bpermute(p, 0) <= 2 And bpermute(p, 1) <= 3 And bpermute(p, 2) <= 3 And bpermute(p, 3) <= 3) _
        And (bpermute(p, 0) + bpermute(p, 1) + bpermute(p, 2) + bpermute(p, 3) <= 8) Then
        'the total number of black pieces that can be chosen from should exceed the minimum piece requirement
        'the maximum piece requirement is irrelevant and will be controlled for automatically later in the code
            For q = 0 To 4
                bpermute_satisfied(bp_counter, q) = bpermute(p, q)
                  'adds the permutation that satisfies the requirement
            Next q
            bp_counter = bp_counter + 1
        End If
    Next p

    If bp_counter > 0 Then
    'assuming at least one permutation that satisfies the requirement was found (quite likely)
        bp_counter = bp_counter - 1 'arrays start at 0
        bp_choice = Generate_Random_Number(0, bp_counter)
        'a random one is selected
        For q = 0 To 4
        'and the values filled in the Composer form
            Composer.BQS.Text = bpermute_satisfied(bp_choice, 0)
            Composer.BRS.Text = bpermute_satisfied(bp_choice, 1)
            Composer.BBS.Text = bpermute_satisfied(bp_choice, 2)
            Composer.BNS.Text = bpermute_satisfied(bp_choice, 3)
            Composer.BPS.Text = bpermute_satisfied(bp_choice, 4)
        Next q
    Else
        total_attempts = total_attempts - 1
        GoTo 1  'or the composing attempt has to start all over again
```

```
    End If
End If

2

Total_Paused_Time = Total_Paused_Time + (Val((DateDiff("s", Pause_Start_Time, Pause_End_Time))))
Pause_Start_Time = Pause_End_Time  'so the pause time is effectively reset to 0

Composer.CTE.Text = Main.Convert_Seconds(Val((DateDiff("s", Composer_Start_Time, Date + Time))) -
Total_Paused_Time)

elapsed_time = Val((DateDiff("s", Composer_Start_Time, Date + Time))) - Total_Paused_Time
'the elapsed time composition has been running

If Composer.ISL.value = 1 Then
'if intelligent switching is selected
    If (elapsed_time / 60) / factor > Generate_Random_Number(720, 1440) Then
    'checks between every 12 to 24 hours (using seconds would exceed the integer limitation of the GRN function)
    'the first time around, the factor is 1 but after say, 12 hours have passed, the factor is now 2, so 24 hours
    should have passed before _ repeating the following
      factor = factor + 1
      'this keeps track of how many cycles of at least 12 hours have passed
      If composing_efficiency > is_ce Then
      'this means efficiency is good
      'so conventions are increased (probably), minimum aesthetics increased and logical filter made more strict
        is_ce = composing_efficiency
        If Generate_Random_Number(1, 8) = 1 Then
          If Composer.No_Duals.value = 0 Then Composer.No_Duals.value = 1
        End If
        If Generate_Random_Number(1, 8) = 2 Then
          If Composer.No_Checks.value = 0 Then Composer.No_Checks.value = 1
        End If
        If Generate_Random_Number(1, 8) = 3 Then
          If Composer.No_Captures.value = 0 Then Composer.No_Captures.value = 1
        End If
        If Generate_Random_Number(1, 8) = 4 Then
          If Composer.No_Restrict.value = 0 Then Composer.No_Restrict.value = 1
        End If
        If Generate_Random_Number(1, 8) = 5 Then
          If Composer.No_Promo.value = 0 Then Composer.No_Promo.value = 1
        End If
        If Generate_Random_Number(1, 8) = 6 Then
          If Composer.MTOL.value = 0 Then Composer.MTOL.value = 1
        End If
        If Generate_Random_Number(1, 8) = 7 Then
          If Composer.DMEL.value = 0 Then Composer.DMEL.value = 1
        End If
        If Generate_Random_Number(1, 8) = 8 Then
          If Composer.NSMP.value = 0 Then Composer.NSMP.value = 1
        End If

        Composer.Min_Aesthetics.Text = Val(Composer.Min_Aesthetics.Text) + 0.1
        Call Logical_Filter_Handler("strict")
      Else
      'this means efficiency has dropped
      'so one of the conventions (except for no cooks) is unchecked in the hope of improving efficiency
        If composing_efficiency < is_ce Then
        'if efficiency hasn't changed, do nothing
        'otherwise, efficiency has dropped so lower requirements
          If Generate_Random_Number(1, 4) = 1 Then
            If Composer.No_Duals.value = 1 Then Composer.No_Duals.value = 0
          End If
          If Generate_Random_Number(1, 8) = 2 Then
            If Composer.No_Checks.value = 1 Then Composer.No_Checks.value = 0
```

```
                End If
                If Generate_Random_Number(1, 8) = 3 Then
                   If Composer.No_Captures.value = 1 Then Composer.No_Captures.value = 0
                End If
                If Generate_Random_Number(1, 8) = 4 Then
                   If Composer.No_Restrict.value = 1 Then Composer.No_Restrict.value = 0
                End If
                If Generate_Random_Number(1, 8) = 5 Then
                   If Composer.No_Promo.value = 1 Then Composer.No_Promo.value = 0
                End If
                If Generate_Random_Number(1, 8) = 6 Then
                   If Composer.MTOL.value = 1 Then Composer.MTOL.value = 0
                End If
                If Generate_Random_Number(1, 8) = 7 Then
                   If Composer.DMEL.value = 1 Then Composer.DMEL.value = 0
                End If
                If Generate_Random_Number(1, 8) = 8 Then
                   If Composer.NSMP.value = 1 Then Composer.NSMP.value = 0
                End If

                If Val(Composer.Min_Aesthetics.Text) > 0 Then Composer.Min_Aesthetics.Text =
                     Val(Composer.Min_Aesthetics.Text) - 0.1
                'make sure it cannot drop below 0

                Call Logical_Filter_Handler("lax")
             End If
          End If
       End If
    End If

    If Composer.st.value = 1 Then
    'if switching techniques enabled
    'this happens internally and does not show on the form
       If composing_efficiency < is_ce Then
       'if efficiency has dropped
          If chosen_technique = 3 Then
             chosen_technique = 1
             'Call Debug_Print("Switched from DSNS to Experience because efficiency dropped from " & is_ce & "
               cph" & "to " & _ composing_efficiency & " cph.")

          End If
          'switch between DSNS and experience
          If chosen_technique = 1 Then
             If Excel_DSNS_File_Name <> vbNullString Then
             'if even the first DSNS source file name is not available, that means none was specified initially and
               switching _ techniques cannot take place

               chosen_technique = 3
               'Call Debug_Print("Switched from Experience to DSNS because efficiency dropped from " & is_ce & "
                  cph" & "to " & _ composing_efficiency & " cph.")

             End If
          End If
       Else
          is_ce = composing_efficiency
       End If
    End If

    If DSNS_Experiment = True Or ER_Experiment = True Then
       Composer.Experiment.Caption = "Cycle: " & ij + 1 & "/" & Val(Composer.Cycles.Text)

       If ij < Val(Composer.Cycles.Text) Then
          If Val((DateDiff("s", starting_time, Date + Time))) >= Val(Composer.Composing_Time.Text) Then
             composed_this_cycle = Val(Composer.Mi3s.Text) - composed_thus_far
               composition_efficiency = Round((composed_this_cycle / (Val(Composer.Composing_Time.Text) / 60 / 60)), 2)

             composed_thus_far = Val(Composer.Mi3s.Text)
```

```
            Open app.Path & "\" & "Experimental Results.txt" For Append As #33
            Print #33, "Composing Efficiency, Cycle " & ij + 1 & ": " & composition_efficiency & " cph"  Close #33
            If ij = 0 Then starting_time_original = starting_time 'backs up the original starting time
            starting_time = Date + Time 'resets it
            ij = ij + 1
        End If
    Else
        Composer.Stop_Button_Execute
        'ends the composer
        ending_time = Date + Time
        Open app.Path & "\" & "Experimental Results.txt" For Append As #33
        Print #33,
        Print #33, "Ended: " & Date & vbTab & Time
        Print #33, "Total Time: " & vbTab & Main.Convert_Seconds(Val(DateDiff("s", starting_time_original,ending_time)))
        Close #33
        ij = 0
        DSNS_Experiment = False: ER_Experiment = False: Composer.Pause.Enabled = True
    End If
End  If

For a1 = 1 To 2
'places the white and black king at a random locations on the board
    Do
        b1 = Generate_Random_Number(0, 63)
        DoEvents
    Loop Until Current_Position(b1) = 0
    If a1 = 1 Then
        white_king_location = b1: Current_Position(b1) = 6
    Else
        black_king_location = b1: Current_Position(b1) = 12
    End If
    DoEvents
Next a1

If Chebyshev_Distance(white_king_location, black_king_location) = 1 Then
'a quick test to see if the kings are next to each other, making it an illegal position
    Call Clear_the_Board
    GoTo 2
Else
    If Check_Legal(Current_Position) = True Then
    'even though the position is already legal, this is needed to generate the FEN
        Call RP_Sub
        'automatically reads the position into the main board
    End If
End If

mnuDBFTPFT.Enabled = False  'this disables the forced ftp transfer option once the composing process has been started

c = 0: white_piece = True
Do While c <= d
'this keeps the cycle going for a reasonable amount of time (but not ad infinitum)
    If white_piece = True Then
    'the probability of choosing a blank piece (kings not included, as they were put on the board earlier) is 1 in 6 _
    or 16.67%
    'Black is made twice as likely to have a blank piece (to give White the better likelihood of being able to mate)
        blank_prob = 16.67
    Else
        blank_prob = 33.33
    End If

    If chosen_technique = 3 Then blank_prob = 16.67
    'the DSNS technique does not rely on this (equal for white and black)
```

```
    If Random_Decision(blank_prob) = True Then
    'determine if a piece or a blank will be selected
      white_piece = Not (white_piece): c = c - 1: GoTo 4
    Else
3
      Do
      'a random square (must be unoccupied)
        j = Generate_Random_Number(0, 63)
        DoEvents
      Loop Until Current_Position(j) = 0
      'it is important to place this portion here because when a piece type has been exceeded, there is a chance
to _ place a different piece on a different *square* to avoid arriving at the same position (e.g. after 'transformation')
      If chosen_technique = 1 Then
        n = Generate_Random_Number(1, 5)
        'randomly determine which piece (except kings) will be selected
        '(sequentially running from 1 through 5 makes pawns more likely than the other pieces to be selected)
        If white_piece = False Then n = n + 6
        piece_chosen = n
        'if Black's turn for a piece, adjust the piece code
        If (piece_chosen = 1 And j <= 7) Or (piece_chosen = 7 And j >= 56) Then GoTo 3
        'pawns cannot occupy these ranks

        sqrs_arnd = Squares_Around(j)
        For o = 0 To UBound(sqrs_arnd)
        'the squares immediately around the piece are examined to see if it has a better placement in any of them
          If Current_Position(sqrs_arnd(o)) <> 6 And Current_Position(sqrs_arnd(o)) <> 12 Then
          'kings cannot be replaced by other pieces
            If Current_Position(sqrs_arnd(o)) = 0 Then
            'in the case of an empty surrounding square, it simply ensures that the chosen piece would have a
            higher _ probability placement there than in its existing square 'j'
              If Determine_Probs(sqrs_arnd(o), piece_chosen) > Determine_Probs(j, piece_chosen) Then j = sqrs_arnd(o)
            Else
            'but, if there is an existing piece in a surrounding square, it must have a higher probability than that _
            piece has being on that square
              If white_piece = True Then
              'black pieces cannot replace white pieces
                If Determine_Probs(sqrs_arnd(o), Current_Position(sqrs_arnd(o))) < _
                Determine_Probs(sqrs_arnd(o), piece_chosen) Then
                  j = sqrs_arnd(o)
                End If
              Else
              'only its own
                If White_or_Black_Piece(Current_Position(sqrs_arnd(o))) = "black" And _
                Determine_Probs(sqrs_arnd(o), Current_Position(sqrs_arnd(o))) < _
                Determine_Probs(sqrs_arnd(o), piece_chosen) Then
                  j = sqrs_arnd(o)
                End If
              End If
            End If
          End If
          DoEvents
        Next o
      End If

      If chosen_technique = 2 Or (chosen_technique = 3 And Query_Attribute(0) = 5) Then
      'if only 5 attributes are used for DSNS, this has to be used instead
      '5 attributes only works if both DSNS source files (assuming merging two domains) have 5 attributes
        i = Generate_Random_Number(1, 5)
        piece_chosen = i
        If white_piece = False Then piece_chosen = piece_chosen + 6
      End If
```

```
    If chosen_technique = 3 And Query_Attribute(0) = 10 Then
    'this will work for the DSNS approach (which doesn't use the 'experience' approach)
       Dim r As Integer, s As Integer, t As Integer, u As Integer
       'this part deals with increasing the likelihood of choosing a piece that was either the first or last to move _
       in either of the two DSNS strings presently chosen
       r = Val(dsns_constraints.first_piece_to_move_s1): s = Val(dsns_constraints.last_piece_to_move_s1)
       t = Val(dsns_constraints.first_piece_to_move_s2): u = Val(dsns_constraints.last_piece_to_move_s2)
       i = Generate_Random_Number(1, 5)
       If Piece_On_Internal_Board(r, s, t, u) = False Then
       'if none of these pieces are already on the board
          If white_piece = True Then
             If i <> r And i <> s And i <> t And i <> u Then
             'if the piece does not equal one of the first or last to move in the DSNS strings selected earlier, _
             there is a 50% chance it will be chosen again
                If Random_Decision(50) = True Then
                   i = Generate_Random_Number(1, 5)                'try again
                End If
             End If
          End If
       End If
       piece_chosen = i
       If white_piece = False Then piece_chosen = piece_chosen + 6
    End If
End If
piece_counter_val = 0
For square_counter = 0 To 63
   If Current_Position(square_counter) = piece_chosen Then
      piece_counter_val = piece_counter_val + 1
   End If
Next square_counter
'counts the number of pieces on the board
If pcount_failsafe >= 50 Then GoTo 1
'there a 5 different piece types; if a resolution hasn't been found after 10 cycles of randomly choosing another
piece, it's _ time to quit and start a new composition (because the current position probably can't be improved with the
pieces left to _ choose from)
'otherwise, the composition will tend to get stuck in an infinite loop
Select Case piece_chosen
   Case 5
      If piece_counter_val >= Val(Composer.WQS.Text) Then
         pcount_failsafe = pcount_failsafe + 1: GoTo 3
      End If
   Case 11
      If piece_counter_val >= Val(Composer.BQS.Text) Then
         pcount_failsafe = pcount_failsafe + 1: GoTo 3
      End If
   Case 4
      If piece_counter_val >= Val(Composer.WRS.Text) Then
         pcount_failsafe = pcount_failsafe + 1: GoTo 3
      End If
   Case 10
      If piece_counter_val >= Val(Composer.BRS.Text) Then
         pcount_failsafe = pcount_failsafe + 1: GoTo 3
      End If
   Case 3
      If piece_counter_val >= Val(Composer.WBS.Text) Then
         pcount_failsafe = pcount_failsafe + 1: GoTo 3
      End If
```

```
Case 9
   If piece_counter_val >= Val(Composer.BBS.Text) Then
     pcount_failsafe = pcount_failsafe + 1: GoTo 3
   End If
Case 2
   If piece_counter_val >= Val(Composer.WNS.Text) Then
     pcount_failsafe = pcount_failsafe + 1: GoTo 3
   End If
Case 8
   If piece_counter_val >= Val(Composer.BNS.Text) Then
     pcount_failsafe = pcount_failsafe + 1: GoTo 3
   End If
Case 1
   If piece_counter_val >= Val(Composer.WPS.Text) Then
     pcount_failsafe = pcount_failsafe + 1: GoTo 3
   End If
Case 7
   If piece_counter_val >= Val(Composer.BPS.Text) Then
     pcount_failsafe = pcount_failsafe + 1: GoTo 3
   End If
End Select
'these keep the pieces that are chosen from exceeding that provided in an original piece set (or as desired by
the user) 'should any be exceeded, the counter is reset and a new piece is chosen
pcount_failsafe = 0
'this resets the failsafe counter once a valid piece (i.e. not exceeding any limits) has been chosen

If (piece_chosen = 3 And Val(Composer.WBS.Text) = 2) Or (piece_chosen = 9 And Val(Composer.BBS.Text)
  = 2) Then 'this ensures that bishops, even two, for each army are restricted to opposite-colored squares
  For k = 0 To 63
    If Current_Position(k) = piece_chosen Then
      If White_Sq(k) = White_Sq(j) Then
        GoTo 3
      End If
    End If
    DoEvents
  Next k
End If

If piece_chosen = 3 Or piece_chosen = 9 Then
'in case of bishops
'even though the code above already covers this somewhat, in the DSNS, 3 bishops may be allowed in
WBS/BBS.Text portions so in those _ cases, they should be placed on different color squares
'so only if a third bishop is placed on the board can it be on the same-color square as one of the other two
  If Piece_Count_On_Internal_Board(piece_chosen) = 1 Then
  'if there is already one of that color on the board
    If White_Sq(Piece_Location_On_Internal_Board(piece_chosen)) = White_Sq(j) Then GoTo 3
    'if it is on the same color square as the square chosen (j from way above), just choose another piece and square
  End If
End If
Current_Position(j) = piece_chosen  'this puts the chosen piece on a particular square on the board (determined way above; 'j')
For e = 1 To 4
  If Castling_Is_Possible(e) = True Then
  'determines if any kinds of castling are possible in the position
    If Generate_Random_Number(1, 100) > 50 Then
    'for each possibility of castling, there is set a 50% chance it is registered as possible
    'this 'registration' will remain even after piece reduction or 'optimization' below
      Castling_Type(e - 1).value = 1: castling_flag = True
    End If
  End If
```

```
    DoEvents
  Next e
  'en passant is not accounted for in this composer because it would only be viable to indicate so in the initial _
  position, which potentially gives up the key move (hence not recommended)
  If Check_Legal(Current_Position) = False Then
  'if the piece makes the position illegal for some reason, it is removed
    Current_Position(j) = 0: GoTo 4
  Else
    Call RP_Sub
    If Illegal_Notice.Caption <> "Status OK" Then
    'this also excludes initial positions that have one side in check
      Current_Position(j) = 0: GoTo 4
    Else
      prnm = Piece_Requirements_Not_Met
      'checks the piece count requirements specified in the 'Composer' form
      If prnm = 1 Then GoTo 1
      'exceeded piece limit
      If prnm = 2 Then
        white_piece = Not (white_piece): GoTo 4
      End If
      'if the minimum piece requirement(s) are not met, more pieces are first added
      If chosen_technique = 3 Then
        'this FIRST part deals with conformity to the difference in Shannon value between the armies
        Dim lower_limit, upper_limit As Integer
        Dim white_shannon As Integer, black_shannon As Integer

       If Val(dsns_constraints.shannon_diff_s1) >= Val(dsns_constraints.shannon_diff_s2) Then
          upper_limit = Val(dsns_constraints.shannon_diff_s1): lower_limit = Val(dsns_constraints.shannon_diff_s2)
        Else
          upper_limit = Val(dsns_constraints.shannon_diff_s2): lower_limit = Val(dsns_constraints.shannon_diff_s1)
        End If

        white_shannon = Total_Shannon_Value("white"): black_shannon = Total_Shannon_Value("black")
        If Abs(white_shannon - black_shannon) > upper_limit Then
        'the difference in material between the two armies should not exceed the highest of the two DSNS strings
          GoTo 1
        End If
        'otherwise, a new composition is attempted
        'this SECOND part deals with conformity to the sparsity attributes of the DSNS
        'there's no point measuring sparsity after each piece because clearly with fewer pieces, sparsity will be
        high   so it only makes sense to measure toward the end of the composing process (i.e. here)
        If Query_Attribute(0) = 10 Then
        'if 5 attributes are used, this is not possible

          v = Val(dsns_constraints.sparsity_score_s1): w = Val(dsns_constraints.sparsity_score_s2)

          Current_Position(j) = i
          sparsity_value = Sparsity_Theme("composition")
          Current_Position(j) = 0

          If v + w >= 1 Then
          'together, both present DSNS string sparsity attributes lean toward greater sparsity
          'specifying that the position generated be between the upper and lower limits of the present DSNS
          strings was tried _ but it caused too many to be rejected (e.g. the first one is 0.24 and the second is 0.35
          so the range is too narrow)
            If sparsity_value < 0.25 Then
            'so if the currently composed position is particular dense, it does not resemble the DSNS strings
              GoTo 1
```

```
                End If
            Else
            'both present DSNS string sparsity attributes lean toward greater density
                If sparsity_value > 0.75 Then
                'if particularly sparse, it doesn't resemble the DSNS strings
                    GoTo 1
                End If
            End If
        End If

    End If

    'the block of code in green below can be used to force the position on the composer (use either Goto 11 or 12, not both)
    'Forsythe.Text = "2n4b/1rn2BN1/6P1/q1pP4/4b1K1/2N1Q3/1N2P3/7k w - - 0 1"
    'Call RP_Sub
    'final_mate = True
    'Call Code_View.Delay(10000000)
    'GoTo 11 'three-movers (just one place of quick write to file; otheriwise remove this Goto and let just continue below)
    'GoTo 12 'long mate (there are two places with the option of quick write to file or checking conditions first)
    If mnuOptCompLM.Checked = True Then GoTo 13
    'if the option to compose only long mates was selected
    Call MSUCI(Val(Composer.Solving_Time.Text))  'tests for #3
    Do Until composer_continue = True
    'this allows the mate solver's analysis file to fully write and the 'optimizable' flag to be set before the _
    rest of the code is read
    'otherwise, especially below, the mates may not be properly 'reduced' or optimized
      DoEvents
     Loop

    If optimizable = True Then
        'if so, the following removes one piece at a time to see if the #3 can be made more economical
        'this may cause the piece count requirements not to be met
        'if the user has selected that the position *not* always be optimized economically (to generate more
        complex mates) _ then a 50% chance probability of skipping one round of this optimization block is set
        optimizable = False 'resets the flag

        temp_fen = Forsythe.Text: eco_rounds = 3
6
        temp_gcm = GCM  'the original mating sequence is recorded here, before pieces are successfully
       removed (creating a _new GCM)

        For I = 1 To eco_rounds
        'Reduce the White pieces first, then Black, then White again (2 or 3-phase, depending)
            If I = 1 Or I = 3 Then
                piece_color = "white"
            Else
                piece_color = "black"
            End If
            For m = 0 To 63
                If Current_Position(m) <> 0 And Current_Position(m) <> 6 And Current_Position(m) <> 12 And _
                White_or_Black_Piece(m) = piece_color Then
                'skips blank squares and kings
                    temp_piece = Current_Position(m): Current_Position(m) = 0
                    'makes a temporary copy of the piece to be removed and the (current) generated composition
                     moves _ which is a global array
                    If Check_Legal(Current_Position) = True Then
                    'this is necessary to generate a revised FEN notation in the main form's corresponding textbox
                        GCM = vbNullString
                        'erases the current GCM
                              Call  RP_Sub
```

```
            If Illegal_Notice.Caption <> "Status OK" Then
                GCM = temp_gcm: Current_Position(m) = temp_piece
            Else
                Call MSUCI(Val(Composer.Solving_Time.Text))
                'tests for #3 in the 'optimized' position; the 'Forced Mi3 Line' function (called by MSUCI) _
                ensures that shorter and longer mates do not qualify
                Do Until composer_continue = True
                    DoEvents
                      Loop

                If optimizable = False And convention_violated = True Then
                'if, after reducing the mate position to economical form there is a convention violation,
                more pieces _  are added convention_violated = False

                    GoTo 3
                End If

                If optimizable = False And convention_violated = False Then
                'if there is no #3
                    Current_Position(m) = temp_piece: GCM = temp_gcm
                    'the piece is put back on the board and the original GCM restored
                    'if the cde flag is true here, it does not matter because the piece that _
                    caused the position to produce an error with the mate-solver is removed to _
                    restore the position to its earlier state
                Else
                'if there is, the flag is restored
                    optimizable = False
                End If

                If Check_Legal(Current_Position) = True Then
                'a new (or the old) FEN is generated assuming its legal
                    If GCM <> temp_gcm Then
                    'if removing a piece changes the original mating sequence, that piece is put back
                        Current_Position(m) = temp_piece: GCM = temp_gcm
                    End If

                    If Composer.AO.value = 0 And Random_Decision(50) = True Then
                    'if the 'always economical/optimize' is not checked, the following block tries to ensure
                    that an _ optimized composition is not too different from original unoptimized form at the expense
                    of _  perhaps some economy
                        If eco_rounds > 2 Then
                            Forsythe.Text = temp_fen: GCM = temp_gcm
                            Call RP_Sub
                            'reduces by only one cycle to prevent uneconomical compositions occurring too often
                            eco_rounds = eco_rounds - 1
                            GoTo 6
                        End If
                    End If

                    Call RP_Sub
                    If Illegal_Notice.Caption <> "Status OK" Then
                        GCM = temp_gcm: Current_Position(m) = temp_piece
                    End If
                End If
            End If
        Else
        'if the new position is not legal somehow, the piece is simply put back
        'an updated FEN was never generated so there is no need 'Check_Legal' here
                Current_Position(m) = temp_piece
        End If
    End If
    DoEvents
Next m
```

```
    DoEvents
  Next I

  If Composer.Cplx.value = 1 And Random_Decision(75) = True Then
  'only if the option is checked and 75% chance at that (so most of the time, but not *all* the time)
  'this is because full economization (Composer.AO) sometimes creates interesting compositions
  'this part adds pieces to see if the position can be made more complex
  'the idea is to add a piece (randomly) and if the mate changes (from the optimized _
   version above), that means its complexity has increased
  'testing all the squares or using the experience file around existing pieces caused ridiculous positions
  like with _ 5 queens next to each other (which still had valid mates) so this random approach was used
9
    ReDim gcm_strings(0 To 32) As String
    'the number of GCMs can't be too many, after all only about 10 pieces or so can be added

    gcm_strings(gcm_counter) = GCM
    'so gcm_strings(0) is the original GCM
    'the gcm_counter does not need to be updated except when another valid mate (with an additional
     piece) _ has been successfully placed on the board with mate possible (see way below)

    If chosen_technique = 3 Then
    'if the DSNS composing technique is chosen, the sum year is taken (for a potentially bigger
    additive_value) and based _ on one iteration of the additive number concept, the upper limit for the ptc is determined
    'so if we have two years, 1980 and 2000, the former is chosen and the ptc_limit becomes 18
      sum_year = Val(dsns_constraints.year_s1) + Val(dsns_constraints.year_s2)
      'earlier_year = MinMax(Val(dsns_constraints.year_s1), Val(dsns_constraints.year_s2), "min")
      ptc_limit = CInt(Additive_Result(sum_year, 1))
    Else
    'otherwise, 12 is set as a fair value
    '12 is used instead of 10 to make the possibility of pawns a little more likely (there are more of these in any game)
      ptc_limit = 12
    End If

    For ptc = 1 To ptc_limit
    'a certain number of cycles of piece types are done to give each a fair chance
    'the fact that having more than or equal to two pieces already on the board causes another piece to
     be selected _ (refer below), means pawns will more likely be chosen anyway (no further programming controls
     required) 'there is no inefficiency here by not ruling out that particular piece type once it is placed because the
     position _ may have changed since the last piece (of the same type) or some other type was put on the board

      If ((Calculate_Material("white") - Calculate_Material("black")) > 5) Or _
      ((Piece_Count("white") - Piece_Count("black")) > 2) Then
      'if the value of the white pieces exceeds the black pieces by more than 5 Shannon points (e.g. a
      rook), _ or the number of white pieces exceeds the number of black pieces by three or more, the
      probability of the next _ piece being chosen as white is set to much lower
        mbp = 30
      Else
        mbp = 50
      End If

      If Random_Decision(mbp) = True Then
      'the probability for choosing a white piece is lowered slightly to create a composition that is more _
      balanced in terms of army strength
        pta = Generate_Random_Number(1, 5)  'white piece type selected
      Else
        pta = Generate_Random_Number(7, 11) 'black piece type selected
      End If
```

```
If Piece_On_Internal_Board(pta) = True Then
'the Check_Legal function below ensures the pieces, in summation are all possible and legal (pawns, promotions etc.)
  Select Case pta
    Case 1, 7
      'if there are more than 7 pawns of that color already on the board, then skip
      If Piece_Count_On_Internal_Board(pta) >= 7 Then GoTo 8
    Case 5, 11
      'if there is already a queen of that color, do not add another one
      If Piece_Count_On_Internal_Board(pta) >= 1 Then GoTo 8
    'For bishops
      'this is handled below (after the location of the piece is determined)
      'to avoid two bishops of the same color
    Case Else
      'for all other pieces, if more than 2 already on the board, do not add anymore
      If Piece_Count_On_Internal_Board(pta) >= 2 Then GoTo 8
  End Select
End If

Dim empty_squares() As Integer
Dim empsc As Integer 'empty square counter
Dim sel_emp As Integer 'selected empty square

empsc = 0

ReDim empty_squares(0 To 63) As Integer
'this is necessary here or you can't redim preserve below

For m = 0 To 63
'log all the empty squares in an array
  If Current_Position(m) = 0 Then
    empty_squares(empsc) = m
    empsc = empsc + 1
  End If
Next m

ReDim Preserve empty_squares(0 To empsc - 1) As Integer
'resize the array to contain only the empty squares

If Forsythe.Text = current_fen Then
'if the present position is the same as the last position, this implies that the cycle below has gone _
through completely with no change
'it can happen in, e.g. 8/4K3/8/2R5/8/8/1P6/k7 w - - 0 1 where the white rook is replace by both
black _ and white pieces yet nothing seems to change
  GoTo 4
Else
  current_fen = Forsythe.Text 'logs the current FEN before the new piece is placed
End If

For m = 0 To UBound(empty_squares)
'try as many times as there are empty squares
    Do
  'this ensure the squares are tried at random to avoid a buildup of pieces next to each other
    sel_emp = Generate_Random_Number(0, UBound(empty_squares))
    'selects any one of the slots in the array at random
    DoEvents
  Loop Until empty_squares(sel_emp) <> 99

  placement = empty_squares(sel_emp): empty_squares(sel_emp) = 99
  'notes the location and erases that square from the array (to mark it as already tested)

    If pta = 3 Or pta = 9 Then
  'in case of bishops
    If Piece_Count_On_Internal_Board(pta) = 1 Then
    'if there is already one of that color on the board
```

```
            If White_Sq(Piece_Location_On_Internal_Board(pta)) = White_Sq(placement) Then GoTo
7
            'if it is on the same color square as the square chosen, just choose another square
         End If
      End If

      If ((pta = 1 And placement <= 7) Or (pta = 7 And placement >= 56)) Or _
      ((pta = 1 And (placement >= 8 And placement <= 15)) And Random_Decision(30)) = True Then
      'pawns cannot occupy the back ranks, so do nothing (i.e. choose another square) in these
      cases  'effectively, this is as if Goto 7 were written here
      'the probability of placing a white pawn on the 7th rank is also significantly reduce to minimize _
      the occurrence of mates by promotion to queen (30% to standardize with the above)
      Else
         Total_Paused_Time = Total_Paused_Time + (Val((DateDiff("s", Pause_Start_Time, Pause_End_Time))))
         Pause_Start_Time = Pause_End_Time
         'so the pause time is effectively reset to 0

         Composer.CTE.Text = Main.Convert_Seconds(Val((DateDiff("s", Composer_Start_Time,
         Date + Time))) - _  Total_Paused_Time)

         'updates the time elapsed because this stage could take some time

         If Val((DateDiff("s", Composer_Start_Time, Date + Time))) <> 0 Then
         'to avoid a division by zero error in some cases
            composing_efficiency = RoundIt((Val(Composer.Mi3s.Text) + long_mate_compositions +
            study_compositions) _ / ((Val((DateDiff("s", Composer_Start_Time, Date + Time))) - Total_Paused_Time) / 3600), 2)
         End If
         Composer.Comp_Eff = composing_efficiency & " cph": Composer.Comp_Eff.ToolTipText = "compositions per hour"
         Main.Caption = Mid(Main.Caption, 1, main_cap_length) & " - " & composing_efficiency & "cph" 'composing efficiency
         If Stop_Composing = True Then GoTo 8
         'in case the user wishes to terminate (it will cycle quickly through the 10 ptc iterations)

         'this portion is repeated here is because in experiments, the cycle counter is often not updated in time
         If DSNS_Experiment = True Or ER_Experiment = True Then
            Composer.Experiment.Caption = "Cycle: " & ij + 1 & "/" & Val(Composer.Cycles.Text)
            If ij < Val(Composer.Cycles.Text) Then
               If Val((DateDiff("s", starting_time, Date + Time))) >= Val(Composer.Composing_Time.Text) Then
                  composed_this_cycle = Val(Composer.Mi3s.Text) - composed_thus_far
                  composition_efficiency = Round((composed_this_cycle / (Val(Composer.Composing_Time.Text) / 60 / 60)), 2)
                  composed_thus_far = Val(Composer.Mi3s.Text)

                  Open app.Path & "\" & "Experimental Results.txt" For Append As #33
                  Print #33, "Composing Efficiency, Cycle " & ij + 1 & ": " & composition_efficiency & "cph"
                  Close #33
                  If ij = 0 Then starting_time_original = starting_time 'backs up the original starting time
                  starting_time = Date + Time 'resets it
                  ij = ij + 1
               End If
            Else
               Composer.Stop_Button_Execute
               'ends the composer
                     ending_time = Dat     e + Time
               Open app.Path & "\" & "Experimental Results.txt" For Append As #33
               Print #33,
               Print #33, "Ended: " & Date & vbTab & Time
```

```
        Print #33, "Total Time: " & vbTab & Main.Convert_Seconds(Val(DateDiff("s",
        starting_time_original, ending_time)))
        Close #33
        ij = 0
        DSNS_Experiment = False: ER_Experiment = False: Composer.Pause.Enabled = True
    End If
End If

Current_Position(placement) = pta  'puts the piece on the board

If Check_Legal(Current_Position) = True Then
'this is necessary to generate a revised FEN notation in the main form's corresponding textbox
    GCM = vbNullString
    'erases the current GCM
    Call RP_Sub
    If Illegal_Notice.Caption <> "Status OK" Then
        Current_Position(placement) = 0: GCM = gcm_strings(gcm_counter): GoTo 7
        'if the position is illegal, removes the piece and restores the GCM
        'gcm_counter should be used instead of '0' for the array index because the restoration
        should _ be to the last successful mate based on the piece added

        'should no piece have been added, then the gcm_counter would be zero, pointing to the
        _ original GCM
    Else
        Call MSUCI(Val(Composer.Solving_Time.Text))
        'tests for #3 in the 'optimized' position; the 'Forced Mi3 Line' function (called by MSUCI) _
        ensures that shorter and longer mates do not qualify
        Do Until composer_continue = True
            DoEvents
                Loop

        If optimizable = False And convention_violated = True Then
        'if there is a mate but there is a convention violation, the piece is removed
        ''optimizable' here is a variable reused even for adding pieces in this case
            convention_violated = False
            Current_Position(placement) = 0: GCM = gcm_strings(gcm_counter): GoTo 7
            'there is no 'continue' statement in VB6 so this has to be used
        End If

        If optimizable = False And convention_violated = False Then
        'if there is no #3
            Current_Position(placement) = 0: GCM = gcm_strings(gcm_counter): GoTo 7
        Else
        'if there is, the flag is restored
            optimizable = False
        End If

        For gcm_i = 0 To gcm_counter
            If GCM = gcm_strings(gcm_i) Then
                gcm_match = True: Exit For
            End If
        Next gcm_i

        If gcm_match = False Then
        'so there is a mate, but does it have a different move sequence than the original
        optimized _ one from the piece reduction process earlier on and all the other GCMs with potentially
        added _ pieces (using this part of the program)?
        'if it does not, chances are it's just an uneconomical piece on the board
        'otherwise, it is necessary to the mate
            If Check_Legal(Current_Position) = True Then
            'a new (or the old) FEN is generated assuming its legal
                Call RP_Sub
```

```
                    If Illegal_Notice.Caption <> "Status OK" Then
                        Current_Position(placement) = 0: GCM = gcm_strings(gcm_counter): GoTo 7
                    End If

                    gcm_counter = gcm_counter + 1: gcm_strings(gcm_counter) = GCM:
                    'adds the present successful piece addition GCM produced to the list
                    GoTo 8
                    'once the piece has been placed successfully, no need to keep placing the same _ piece again
                End If
            Else
                gcm_match = False 'resets the flag
                Current_Position(placement) = 0: GCM = gcm_strings(gcm_counter): GoTo 7
                'it's the same mating sequence as the optimized one so no improvement
            End If
        End If
    Else
    'if the new position is not legal somehow, the piece is simply put back
    'an updated FEN was never generated so there is no need 'Check_Legal' here
        Current_Position(placement) = 0: GCM = gcm_strings(gcm_counter): GoTo 7
    End If
End If
DoEvents
7
    Next m
8
Next ptc
Erase gcm_strings: gcm_counter = 0

swpc = 0

ReDim swp(0 To 63) As Integer

For x = 0 To 63
'log all the squares that contain pieces
    If Current_Position(x) >= 1 And Current_Position(x) <= 11 Then
    'all pieces except kings
        swp(swpc) = x
        swpc = swpc + 1
    End If
Next x

ReDim Preserve swp(0 To swpc - 1) As Integer
'this is so that both white and black pieces can be removed systematically (at random) to make
economical _the position with added pieces above (somehow, the order of adding pieces renders the failsafes
above sometimes _ useless so another round of economy optimizing is necessary)... but random this time
original_gcm = GCM

For m = 0 To UBound(swp)
        Do
    'this ensure the squares are tried at random to avoid a buildup of pieces next to each other
        sel_pl = Generate_Random_Number(0, UBound(swp))
        'selects any one of the slots in the array at random
        DoEvents
    Loop Until swp(sel_pl) <> 99

    ptr = swp(sel_pl): swp(sel_pl) = 99
    'notes the piece to remove and tags the array slot as having been tested already
    Total_Paused_Time = Total_Paused_Time + (Val((DateDiff("s", Pause_Start_Time, Pause_End_Time))))
    Pause_Start_Time = Pause_End_Time
    'so the pause time is effectively reset to 0
```

```
Composer.CTE.Text = Main.Convert_Seconds(Val((DateDiff("s", Composer_Start_Time, Date +
Time))) - _ Total_Paused_Time)
'updates the time elapsed because this stage could take some time

If Stop_Composing = True Then GoTo 5
'in case the user wishes to terminate (it will cycle quickly through the 10 ptc iterations)

temp_piece = Current_Position(ptr): Current_Position(ptr) = 0
'makes a temporary copy of the piece to be removed and the (current) generated composition moves

If Check_Legal(Current_Position) = True Then
'this is necessary to generate a revised FEN notation in the main form's corresponding textbox
  GCM = vbNullString
  'erases the current GCM
  Call RP_Sub
  If Illegal_Notice.Caption <> "Status OK" Then
    GCM = original_gcm: Current_Position(ptr) = temp_piece
  Else
    Call MSUCI(Val(Composer.Solving_Time.Text))
    'tests for #3 in the 'optimized' position; the 'Forced Mi3 Line' function (called by MSUCI) _
    ensures that shorter and longer mates do not qualify
    Do Until composer_continue = True
      DoEvents
       Loop
    If optimizable = False And convention_violated = True Then
    'if there is a mate but there is a convention violation, the piece is removed
    ''optimizable' here is a variable reused even for adding pieces in this case
      convention_violated = False
      Current_Position(ptr) = temp_piece: GCM = original_gcm
      'there is no 'continue' statement in VB6 so this has to be used
    End If

    If optimizable = False And convention_violated = False Then
    'if there is no #3
      Current_Position(ptr) = temp_piece: GCM = original_gcm
    Else
    'if there is, the flag is restored
      optimizable = False
    End If

    If Check_Legal(Current_Position) = True Then
    'a new (or the old) FEN is generated assuming its legal
      If GCM <> original_gcm Then
      'if removing a piece changes the original mating sequence, that piece is put back
      'it is difficult to say for certain what the 'original' mating sequence was looking at the final _
      composed position... it may be that the final position looks like it can be made more
      economical _ by removing a piece but the order in which those pieces were placed is not know which _
      is why it may look like this code here is not working well; sometimes the additional piece plays a _
      role in a variation of the mate so the piece is not completely useless as a chess construct, at least
      'in general, if most of the time the resulting composition is economical, it shows the code here _
      and similar code above is probably working
        Current_Position(ptr) = temp_piece: GCM = original_gcm
      End If
      Call RP_Sub
      If Illegal_Notice.Caption <> "Status OK" Then
        Current_Position(ptr) = temp_piece: GCM = original_gcm
      End If
    End If
  End If
Else
'if the new position is not legal somehow, the piece is simply put back
```

```
    'an updated FEN was never generated so there is no need 'Check_Legal' here Current_Position(ptr) = temp_piece
    End If
    DoEvents
  Next m

  If Val((DateDiff("s", Composer_Start_Time, Date + Time))) <> 0 Then
    composing_efficiency = RoundIt((Val(Composer.Mi3s.Text) + long_mate_compositions +
    study_compositions) _ / ((Val((DateDiff("s", Composer_Start_Time, Date + Time))) - Total_Paused_Time) / 3600), 2)
  End If
  Composer.Comp_Eff = composing_efficiency & " cph": Composer.Comp_Eff.ToolTipText = "compositions per hour"
  Main.Caption = Mid(Main.Caption, 1, main_cap_length) & " - " & composing_efficiency & " cph" 'composing efficiency
  'this portion is repeated here is because in experiments, the cycle counter is often not updated in time
  If DSNS_Experiment = True Or ER_Experiment = True Then
    Composer.Experiment.Caption = "Cycle: " & ij + 1 & "/" & Val(Composer.Cycles.Text)
    If ij < Val(Composer.Cycles.Text) Then
      If Val((DateDiff("s", starting_time, Date + Time))) >= Val(Composer.Composing_Time.Text) Then
        composed_this_cycle = Val(Composer.Mi3s.Text) - composed_thus_far
        composition_efficiency = Round((composed_this_cycle / (Val(Composer.Composing_Time.Text) / 60 / 60)), 2)
        composed_thus_far = Val(Composer.Mi3s.Text)

        Open app.Path & "\" & "Experimental Results.txt" For Append As #33
        Print #33, "Composing Efficiency, Cycle " & ij + 1 & ": " & composition_efficiency & " cph"
        Close #33
        If ij = 0 Then starting_time_original = starting_time 'backs up the original starting time
        starting_time = Date + Time 'resets it
        ij = ij + 1
      End If
    Else
      Composer.Stop_Button_Execute
      'ends the composer
        end ing_time = Date + Time
      Open app.Path & "\" & "Experimental Results.txt" For Append As #33
      Print #33,
      Print #33, "Ended: " & Date & vbTab & Time
      Print #33, "Total Time: " & vbTab & Main.Convert_Seconds(Val(DateDiff("s", starting_time_original, ending_time)))
      Close #33
      ij = 0
      DSNS_Experiment = False: ER_Experiment = False: Composer.Pause.Enabled = True
    End If
  End If

  If ((Piece_Count("white") - Piece_Count("black")) > 2) Or Piece_Count("black") = 1 Then
    If repeat_count < ptc_limit Then
      repeat_count = repeat_count + 1: GoTo 9
    End If
  Else
    repeat_count = 0
  End If
  'this reduces the frequent occurrence of a sole black king against several white pieces and _
  also reduces occurrences where white is significantly greater than black in material (three or more pieces than black)
  'it will keep trying until it successfully manages to add at least another black piece without compromising
  the _ economy of the mate.
  'unfortunately, in some positions (e.g. 8/4k3/6K1/3Q4/8/8/8/8             w - - 0 1), this never ends.
  'so the solution is to keep trying to a point (e.g. ptc_limit)
End If
```

```
'the following two lines 'registers' the final optimized/unoptimized position after the cycle above
GCM = vbNullString
Call RP_Sub

final_mate = True
'signifies that the position is 'reduced' enough to be tested for comformity to conventions

If Main.mnuOptCompTM.Checked = False Then GoTo 13
'if three-movers are not to be composed, then jump straight to the long mate section

Call MSUCI(Val(Composer.Solving_Time.Text))
Do Until composer_continue = True
  DoEvents
 Loop

If GCM <> vbNullString Then
'mates shorter or longer than 3 do not produce any GCM (erased in Forced_Mi3_Line subroutine)
'this is to keep their FENs from being written to file in any case

   If Composer.No_Restrict.value = 1 Then
   'ensures that the first move does not restrict the enemy king's move (a composition convention)

      kpm_initial = 0: kpm_after_m1 = 0: temp_gcm = GCM
      Open app.Path & "\" & "Temp_PGNC.pgn" For Output As #30
      Call Register_File_for_Delete_Menu(app.Path & "\" & "Temp_PGNC.pgn", True)
      Call Write_Temp_PGN(30, Forsythe.Text, GCM): Close #30
      Call Delay_File_Access("exists", True, app.Path & "\" & "Temp_PGNC.pgn")
      Call Clear_the_Board
      Call Open_the_Database(app.Path & "\" & "Temp_PGNC.pgn")

      Call PlayBack
      Call Reset_P

      kpm_initial = UBound(Generate_Legal_Moves(12, black_king_location, False, "standard")) + 1
      'the actual number of legal moves ('ubound' is the upper bound of an array; so a 'ubound' of 2
      means _ there are 3 elements, i.e. 0, 1 and 2)

      'having no legal moves still returns an array with one element (i.e. slot 0 with a '0' element _
      inside it); so in such a case the 'no_legal_moves' flag has to be used instead

      If no_legal_moves = False Then
      'the black king has at least one legal move in the initial position, so test after White's first move _
      now
         Call Replay_Move
         kpm_after_m1 = UBound(Generate_Legal_Moves(12, black_king_location, False, "standard")) + 1
         If no_legal_moves = True Then kpm_after_m1 = 0
         'the 'no_legal_moves' flag now applies to kpm_after_m1
            If kpm_after_m1 < kpm_initial Then
         'if White's move reduced the number of legal moves by the black king, the convention is violated
            Current_Position(j) = 0: GoTo 3
         End If
      Else
      'if, in the initial position, the black king cannot move anywhere, no move by White could _
      restrict it further; hence the convention is not violated
         no_legal_moves = False
      End If

      GCM = temp_gcm: Remove_File_Safely (app.Path & "\" & "Temp_PGNC.pgn")

      PGN_Path.Text = vbNullString: Scores_Display.Text = "Score"
      DB_Position.Text = vbNullString: Moves_Display.Text = "Moves"
   End If

   If Composer.MTOL.value = 1 Then
   'if there is only one matin         g line with three moves
```

```
        If NVPC3c < 2 Then
            Current_Position(j) = 0: GoTo 3
            'no need for the convention_violated flag here because action is taken immediately
        End If
    End If

    If convention_violated = True Then
    'for some of the other conventions (e.g. No_Promo)
        convention_violated = False: Current_Position(j) = 0: GoTo 3
    End If

    If Exclusion(Forsythe.Text) = True Then
    'if the optimized position is of the type that should be excluded, the previous piece is removed and
    the process _ of composing continues
        Current_Position(j) = 0: GoTo 3
    End If

    'if intelligent selection is checked, then minimum aesthetics applies (defaults to '0' if nothing entered)

    aesthetics_after_optimization = Get_Aesthetics_Score

    If Composer.ISL.value = 1 Then

        If NVPC <= 60 Then
            discounted_aesthetics = discounted_aesthetics + ((60 - NVPC) / 10)
            '60 is the limit to the number of lines in the YouTube chess problem videos (30 on each side) for
            one frame  'so if there are say, just 20 variation lines (a good thing because the position is more elegant
            and less complicated), _ the discounted aesthetics is 0.4
            'so there is an 'incentive' toward more elegant compositions
            'this will likely also result in more compositions generated
        Else
            discounted_aesthetics = 0
        End If

        If (NVPC - NVPC3) > 0 Then
        'not all the variations are three-movers (i.e. there are some mate-in-2s assuming Black plays poorly)
        'otherwise, division by zero error
            If (NVPC3 / (NVPC - NVPC3)) >= 5 Then
            'if the ratio of three-mover lines to the rest (i.e. two-movers) is more than 5:1 (analysis of previously _
            selected compositions from those generated reveal about this ratio), then add to discounted aesthetics
                discounted_aesthetics = discounted_aesthetics + ((NVPC3 / (NVPC - NVPC3)) / 10)
                'so if the ratio is 6 to 1 or 5.5 to 1, the discounted aesthetics is 0.6 or 0.55
                'an incentive for compositions with more three-mover lines than two-movers (given imperfect play by Black)
            Else
                discounted_aesthetics = 0
            End If
        End If

        If Composer.Logical_Filter.Text <> " " And Composer.Logical_Filter.Text <> vbNullString Then
            If ((NVPC - NVPC3) > 0) Then
                If (NVPC <= 60 And (NVPC3 / (NVPC - NVPC3)) >= 5) Then
                'this provides a temporary exception to the logical filter if the composition has 60 or fewer lines AND _
                at least 5 times more three-mover lines than two-movers
                    suspend_logical_filter = True
                Else
                    suspend_logical_filter = False
                End If
            End If
        End If
```

```
End If

If suspend_logical_filter = False Then
   If Composer.Logical_Filter.Text <> " " And Composer.Logical_Filter.Text <> vbNullString Then
      If PGN_Data.Pieces_and_Material(Forsythe.Text, PGN_Data.PGN_to_Logical(Composer.Logical_
      Filter.Text)) = _ False Then
      'if a logical filter such as bm > wm (black material more than white material) is applied
      'remove the current piece and keep trying
      'this creates positions where white is generally not better off materially than black (more complex mates)
      'random_decision probability here to ensure there is a chance to break out of an otherwise infinite cycle
         Current_Position(j) = 0: GoTo 3
      End If
   End If
End If

If (aesthetics_after_optimization + discounted_aesthetics) < Val(Composer.Min_Aesthetics.Text) Then
'rejects compositions below a specified aesthetics score
   Current_Position(j) = 0: GoTo 3
End If

If Duplicate_Check(Forsythe.Text, file_name, True) = True Then
'checks for duplicate compositions
   Current_Position(j) = 0: GoTo 3
End If

If Composer.CMAL.value = 1 Then GCM = DBVGCM
'this is the point where Chesthetica replaces the main line the external solving engine found (often
not the most _ attractive one) and replaces it instead with the best variation from all the ones the engine found
'the DBVGCM is determined earlier during the aesthetics scoring of the generated composition

11          'Goto 11 here is for quick writing three-mover to file

If GCM = vbNullString Then
'sometimes, for whatever reason, there is no mating line produced
   Current_Position(j) = 0: GoTo 3
End If

If Filexists(file_name) = True Then
   Open file_name For Append As #29
Else
   Open file_name For Output As #29
End If

Call Write_Composition_PGN(29, Forsythe.Text, GCM)
Close #29
Call Register_File_for_Delete_Menu(app.Path & "\" & "Generated Compositions.pgn", True)

Composer.Last_Comp = Date & " " & Time
'logs the date and time of the last composition generated (on screen)

Call Transfer_to_FTP("Generated Compositions")
'this transfers to FTP

If FTP_Reset = True Then
'this means an internal backup of the PGN file was performed
   Mi3s_in_PGN = 0: comp_success = 0: composing_efficiency = 0: total_attempts = 0
   Composer.Total_Mi3s.Text = vbNullString: Composer.Mi3s.Text = vbNullString
   Composer_Start_Time = Date + Time: FTP_Reset = False
   GoTo 11
   'the problem that was just composed needs to be written to file again because after the FTP transfer _
```

```
            (in which it was also written), the whole PGN is backed up and a new one created automatically...
            so the new _ problem needs to be put into that (which will be uploaded the next time around)
          End If

          comp_success = comp_success + 1: Composer.Mi3s = comp_success
          'the number of successfull compositions for this session

          If Val((DateDiff("s", Composer_Start_Time, Date + Time))) <> 0 Then
            composing_efficiency = RoundIt((Val(Composer.Mi3s.Text) + long_mate_compositions + study_compositions) _
            / ((Val((DateDiff("s", Composer_Start_Time, Date + Time))) - Total_Paused_Time) / 3600), 2)
          End If

          Composer.Comp_Eff = composing_efficiency & " cph"
          Composer.Comp_Eff.ToolTipText = "compositions per hour"
          Main.Caption = Mid(Main.Caption, 1, main_cap_length) & " - " & composing_efficiency & " cph"
          'composing efficiency
          Composer.Total_Mi3s = Mi3s_in_PGN + comp_success
          'the total number (including those from an existing PGN file)
          GCM = vbNullString
          'resets and repeats the process for a new generated combination
        End If

        GoTo 1
      Else
      'if there is no #3, add a piece of the opposite color to the board instead

        If cde = True Then
        'if the position is not optimizable and furthermore produced an error with the mate-solver, _
        remove the previous piece
          Current_Position(j) = 0: GoTo 4
        End If

13
        'test to see if there are any longer mates (if composing longer mates is selected)
        If Composer.LFLM.value = 1 Then
        'if the composer only three-movers option is selected, this cannot be checked

'12      'Goto 12 here is for writing long mate to file but with all the checks first

            Call RP_Sub

          LML = Longer_Mate_Exists

          If LML > 3 Then
          'this tests if a longer mates exists in the position and writes it to file

            If Val(Composer.Min_Aesthetics) > 0 Then
            'if a minimum aesthetics value is specified, then Chesthetica Endgame is invoked to calculate the_longer mate 's value
              long_mate_aesthetics = Long_Mate_Aesthetics_Score(Long_Mate_Sequence, Forsythe.Text)

              'Debug.Print "Long Mate: " & long_mate_aesthetics

              If long_mate_aesthetics = 99 Or (long_mate_aesthetics < Val(Composer.Min_Aesthetics)) Then
              'an error or less than specified aesthetics thresh          old
                GoTo 10
              End If
              'calculates the aesthetics using CEG and tests if it is less than the specified threshold
            End If
            For    sythe.Text = Optimize_Long_Mate(Long_Mate_Sequence, LML)
            'this tries to optimize the long mate so that unnecessary pieces can be remove whilst retaining the_
```

```
essential position and the original mating line
Call RP_Sub
If Composer.Logical_Filter.Text <> " " And Composer.Logical_Filter.Text <> vbNullString Then
   If PGN_Data.Pieces_and_Material(Forsythe.Text, PGN_Data.PGN_to_Logical(Composer.
   Logical_Filter.Text)) = False Then
   'if a logical filter such as bm > wm (black material more than white material) is applied keep trying
   'this creates positions where white is generally not better off materially than black (more complex mates)
      GoTo 10
   End If
End If

If Long_Mate_Conventions_Violated(LML) = True Then GoTo 10
If DTrue) = True Then uplicate_Check(Forsythe.Text, app.Path & "\" & "Generated Compositions (Long Mates).pgn",
'checks for duplicate compositions
   GoTo 10
End If

If Exclusion(Forsythe.Text) = True Then GoTo 10
'if the position is of the type that should be excluded

If Composer.CMAL.value = 1 Then
'if choosing the most aesthetic line is selected
   most_aesthetic_lm_line = MALML(Forsythe.Text, LML)
      Do
   DoEvents
   Loop Until malml_complete = True

   If most_aesthetic_lm_line = vbNullString Then GoTo 10
   'sometimes, for whatever reason this can happen

   If InStr(most_aesthetic_lm_line, LML & ". ") = 0 Then GoTo 10
   'sometimes, for whatever reason, the mating line chosen is not the same length as the maximum possible

   'so, for example, for a mate in 5, a mate in 4 line is shown
End If

If Long_Mate_Sequence = vbNullString Then GoTo 10
'sometimes, for whatever reason, this can happen

If InStr(Long_Mate_Sequence, LML & ". ") = 0 Then GoTo 10
'see similar just above

12    'Goto 12 here is for quick writing long mate to file (and also FTP long mate reset; see below)

   If Filexists(app.Path & "\" & "Generated Compositions (Long Mates).pgn") = True Then
   Open app.Path & "\" & "Generated Compositions (Long Mates).pgn" For Append As #55
Else
      Op en app.Path & "\" & "Generated Compositions (Long Mates).pgn" For Output As #55
End If

If Composer.CMAL.value = 1 Then
   Call Write_Compositi          on_PGN(55, Forsythe.Text, most_aesthetic_lm_line)
Else
   Call Write_Composition_PGN(55, Forsythe.Text, Long_Mate_Sequence)
End If
   Clo se #55
Call Register_File_for_Delete_Menu(app.Path & "\" & "Generated Compositions (Long Mates).pgn", True)
Composer.LMC.Text = Val(Composer.LMC.Text) + 1
If Composer.LMC_Old.Text <> vbNullString Then
```

```
        'if there was a pre-existing number of mates composed prior to this, it's value is updated as well
          Composer.LMC_Old.Text = Val(Composer.LMC_Old.Text) + 1
        End If

        Composer.Last_Comp = Date & " " & Time
        'logs the date and time of the last composition generated (on screen)

        Call Transfer_to_FTP("Generated Compositions (Long Mates)")
        'this transfers to FTP

        long_mate_compositions = long_mate_compositions + 1
        'the value in the LMC textbox cannot be used because if there are mates from a previous run, _
        it will mess up the efficiency value (very high because the time just started)
        'it also needs to be static because the composer can be stopped and started again

        If FTP_LM_Reset = True Then
        'that means an internal backup of the PGN file was performed so a simple reset of the long mate counter _ is done
          Composer.LMC.Text = vbNullString: Composer.LMC_Old = vbNullString
          'it's as if a new composing session had started
          FTP_LM_Reset = False
            Go To 12
          'an explanation for this is above (with FTP_Reset)
        End If

        Call Remove_File_Safely(app.Path & "\" & "adjusted_CHESTUCI_Analysis.txt")
        'this adjusted moves file is removed here (in case any processing needs to be done on it)

        GoTo 1
        'if the long mate is successfully composed, the composing process can start over
      End If
    End If

    If Composer.CStudies.value = 1 Then
    'if composing studies has been selected
16
      forsythe_backup = Forsythe.Text
      temp_study_line = Study_Lines(Forsythe.Text, 20, 10000)
      'gets the analysis lines given the position
      'defaults to depth 20 and 10 seconds (or 10,000 miliseconds)

      arlines_study = Split(temp_study_line, vbNewLine)
      'breaks them into individual lines

      If InStr(arlines_study(0), "1. ") = 0 Or InStr(arlines_study(0), "3. ") = 0 Then GoTo 14
      'the houdini engine probably had an internal buffer conflict or something and output just a partial line
      'or the best line may be just two moves, so that's too short

      If InStr(arlines_study(0), "mate") <> 0 Then GoTo 14
      'there is a mate (perhaps longer than five moves or less than 3) so this disqualifies the position
      'checks the first line

      If InStr(arlines_study(UBound(arlines_study) - 1), "black mates") <> 0 Then GoTo 14
      'this means that black mates somehow (checks the last line)

      cp_value = Val(Mid(arlines_study(0), InStr(arlines_study(0), "cp") + 3, (Len(arlines_study(0)) - InStr(arlines_study(0), "cp"))))
      'extracts just the centipawn value (100 centipawns = 1 pawn)
      temp_study_line = Mid(arlines_study(0), 1, (InStr(arlines_study(0), "[")) - 3)
      'this removes the [cp...] part, leaving just the moves

      algebraic_line = Convert_Coordinate_Moves_to_Algebraic(temp_study_line)

      Forsythe.Text = forsythe_backup: Call RP_Sub

      If Sequence_Length(temp_study_line) <> Sequence_Length(algebraic_line) Then GoTo 16
```

```
'due to the complexity of invoked functions and modules, sometimes, the conversion to algebraic is
not done _ properly and needs to be redone until the lengths match

algebraic_line = algebraic_line & " 1-0"
'this is added so it can be more easily determined if the last ply is by Black

If InStrRev(algebraic_line, "1-0") - InStrRev(algebraic_line, ". ") > 7 Then
'this removes black's move if he moves last (even though he loses) because CEG can only process
if White moves last
'the Houdini engine creates the full move list so this removal has to be done here
   algebraic_line = RTrim(Mid(algebraic_line, 1, InStr(algebraic_line, " 1-0")))
   algebraic_line = RTrim(Mid(algebraic_line, 1, InStrRev(algebraic_line, " ")))
Else
   'if the last move is by white, then the line is restored as it was without the 1-0
   algebraic_line = RTrim(Mid(algebraic_line, 1, InStr(algebraic_line, " 1-0")))
End If

If cp_value < 400 Then GoTo 14
'if there isn't a distinct advantage of at least a minor piece + pawn in value, _
it is not considered a worthy "won" study

If Composer.Logical_Filter.Text <> " " And Composer.Logical_Filter.Text <> vbNullString Then
'to ensure it's blank
   If PGN_Data.Pieces_and_Material(Forsythe.Text, PGN_Data.PGN_to_Logical(Composer.Logical_
     Filter.Text)) = False Then
   'if a logical filter such as bm > wm (black material more than white material) is applied keep trying
   'this creates positions where white is generally not better off materially than black (more complex mates)
      GoTo 14
   End If
End If

If Study_Conventions_Violated(algebraic_line, arlines_study) = True Then GoTo 14
'for studies, it checks, at most, for checks and captures in the key move, and promotion to queen moves
'and also no cooks, but in a slightly different way than with mates

Erase arlines_study

If Val(Composer.Min_Aesthetics) > 0 Then
'if a minimum aesthetics value is specified, then Chesthetica Endgame is invoked to calculate the _
study 's value
   'this quickly prints the temporary game info in case CEG cannot process it (for review)

   study_aesthetics = Study_Aesthetics_Score(algebraic_line, Forsythe.Text)

   If study_aesthetics = 99 Or (study_aesthetics < Val(Composer.Min_Aesthetics)) Then
   'an error or less than specified aesthetics threshold
      GoTo 14
   End If
   'calculates the aesthetics using CEG and tests if it is less than the specified threshold
End If

'Debug.Print Forsythe.Text: Debug.Print algebraic_line & "  " & "+" & Round(cp_value / 100, 1): Debug.Print
15
If Filexists(app.Path & "\" & "Generated Compositions (Studies).pgn") = True Then
   Open app.Path & "\" & "Generated Compositions (Studies).pgn" For Append As #64
Else
   Open app.Path & "\" & "Generated Compositions (Studies).pgn" For Output As #64
End If

Call Write_Composition_PGN(64, Forsythe.Text, algebraic_line, "study")
Close #64
Call Register_File_for_Delete_Menu(app.Path & "\" & "Generated Compositions (Studies).pgn", True)
Composer.StudC.Text = Val(Composer.StudC.Text) + 1
If Composer.StudC_Old.Text <> vbNullString Then
'if there was a pre-existing number of mates composed prior to this, it's value is updated as well
   Composer.StudC_Old.Text = Val(Composer.StudC_Old.Text) + 1
End If
```

```
                Composer.Last_Comp = Date & " " & Time
                'logs the date and time of the last composition generated (on screen)

                Call Transfer_to_FTP("Generated Compositions (Studies)")
                'this transfers to FTP

                study_compositions = study_compositions + 1

                If FTP_STDY_Reset = True Then
                'that means an internal backup of the PGN file was performed so a simple reset of the long mate counter _ is done
                  Composer.StudC.Text = vbNullString: Composer.StudC_Old = vbNullString
                  'it's as if a new composing session had started
                  FTP_STDY_Reset = False
                  GoTo 15
                  'an explanation for this is above (with FTP_Reset)
                End If

                GoTo 1
                'if the study is successfully composed, the composing process can start over

              End If
14
              If convention_violated = True Then
              'this removes the last piece and attempts to add more pieces to a non-reduced mate position where
              convention(s) have been _ violated
                convention_violated = False: Current_Position(j) = 0: GoTo 3
              End If

              If Calculate_Material("difference") = 0 And Random_Decision(75) = True Then
              'the position is inversed and pieces added (so there is some variation);
              'but only in cases where the material is the same and 75% chance probability occurred _
              (so exceptions are possible)
              'this also allows variations in the piece count when switching sides
              'the GCM is erased because the move list is now invalid
                Comp_Mirror = True  'a flag so that the FEN does not have black to move (as in typical mirroring)
                Call Mirror_Position
                GCM = vbNullString
                Comp_Mirror = False
                'Call Swap_WB_Specs (may or may not be necessary)
                'these specifications (determined by the DSNS in this case) should also be swapped when the position is _ mirrored

                GoTo 3
              End If
10
              'otherwise, add a piece of the opposite color
              white_piece = Not (white_piece)
            End If
          End If
4
          c = c + 1

          If Stop_Composing = True Then
          'the stop button was clicked
5
            Composer.Stop_Button.Caption = "Start"
            'this makes the button indicate the process has been stopped only when it really has
            total_attempts = 0: c = 0: comp_success = 0
            Composer.Comp_Eff.ToolTipText = vbNullString
            Call Remove_File_Safely(app.Path & "\" & "fen_temp.txt")
            Call Remove_File_Safely(app.Path & "\" & "CHESTUCI_Analysis.txt")

            Main.Caption = "Chesthetica " & About.lblVersion.Caption
            Call Update_Tray_Tooltip: Exit Sub
          End If
        End If
DoEvents
Loop
GoTo 1
End Sub
```

Index

A. Iqbal et al., *The Digital Synaptic Neural Substrate*,
SpringerBriefs in Cognitive Computation, DOI 10.1007/978-3-319-28079-0